CALIFORNIA NATURAL HISTORY GUIDES

FIELD GUIDE TO BIRDS OF THE NORTHERN CALIFORNIA COAST

California Natural History Guides

Phyllis M. Faber, General Editor

Field Guide to BIRDS of the Northern California Coast

Rich Stallcup and Jules Evens
Graphites by Keith Hansen

UNIVERSITY OF CALIFORNIA PRESS

University of California Press, one of the most distinguished university presses in the United States, enriches lives around the world by advancing scholarship in the humanities, social sciences, and natural sciences. Its activities are supported by the UC Press Foundation and by philanthropic contributions from individuals and institutions. For more information, visit www.ucpress.edu.

California Natural History Guide Series No. 109

University of California Press

Cataloging-in-Publication Data on file at the Library of Congress

ISBN 978-0-520-27616-1 (cloth)
ISBN 978-0-520-27617-8 (paper)
ISBN 978-0-520-95838-8 (ebook)

Manufactured in China

22 21 20 19 18 17 16 15 14
10 9 8 7 6 5 4 3 2 1

The paper used in this publication meets the minimum requirements of ANSI/NISO Z39.48-1992 (R 2002) (*Permanence of Paper*). ♾

Graphite illustrations by Keith Hansen, http://keithhansen.com
Cartography and figures by John Culp

Cover illustration: Sanderlings *(Calidris alba)* at Drake's Beach, Point Reyes National Seashore. 2 March 2013. Photograph by Jules Evens.

The publisher gratefully acknowledges the generous contributions to this book provided by the Gordon and Betty Moore Fund in Environmental Sciences.

To my mentor, field companion, and bird brother

Rich Stallcup (1944–2012)

for fostering respect and love of the natural world
for his lifelong contributions to field ornithology
for his humor, friendship, and enduring spirit

–JE, December 2012

You can observe a lot by watching.

—YOGI BERRA

CONTENTS

Color plates follow page 274.

PREFACE

The authors have, between us, nearly a century of experience birding the Northern California coast—nearly 60 years for Rich Stallcup (he started really early!), almost 40 for Jules Evens. Our close friendship spanned most of Evens's birding career, and he acknowledges that Rich shared his vast knowledge of and deep respect for California's birdlife generously over those decades. Both of us were nurtured first by our parents, our early mentors, and by fellow birders met along the trail, but also by bird books that we devoured as young naturalists. Especially helpful was Ralph Hoffman's classic *Birds of the Pacific States* (1927, illustrations by Allan Brooks), along with the early Peterson field guides, and more academic works such as Joe Grinnell and Alden Miller's *The Distribution of the Birds of California* (1944).

In this volume, we have tried to live up to the standards of those earlier works, providing the accuracy of Grinnell and Miller while conveying the beauty and magic of birds presented so artfully by Ralph Hoffman, Roger Tory Peterson, Allan Brooks, and those many ornithologists and bird artists who followed.

Early on in the pursuit of birds, the curious naturalist discovers that he or she has embarked on a lifelong journey. The closer you look, the more is revealed. The more alertly you listen, the subtler the sounds you hear. Identification of species is as good a place to start as any—perhaps the best place. When you first investigate the variation among (and within) species, nature's brilliant diversity, as well as many similarities, start to come into focus. Some species are so similar in appearance, they can be separated only by voice. In others, the males and females are so different in plumage that they appear to be separate species. Bird behavior, too, adds a layer of complexity that enriches the experience. Each species has its own behavioral traits, and those often offer clues to identity. Compare the flight patterns of ravens and crows. Notice the migratory timing of various species of shorebirds. Does that duck you are observing tend to dabble or dive when foraging?

This book is not a typical field guide. A plethora of those are available, and any of the more popular ones covering the western states—Peterson's, National Geographic, Kauffman, or Sibley—are outstanding and will provide the identification tips needed to inform the curious naturalist.

Nor is this book a primer on bird-watching. There are several books available that cover that subject admirably, and it would be redundant to repeat that information here. *Birding* (2007) by Joe Forshaw, Steve Howell, Terence Lindsey, and Rich Stallcup (Nature Company Guides) and *Sibley's Birding Basics* (2002) by David Allen Sibley (Alfred Knopf) are two recommended sources. Also, the companion to this volume, *Introduction to Birds of the Southern California Coast* (2005) by Joan Easton Lentz (California Natural History Guides, University of California Press) offers an excellent overview of general guidelines to follow when trying to figure out which bird is what.

This book, then, is an introduction to those species that are most likely to be encountered in, or that are most representative of, coastal Northern California. We have written species accounts that try to capture the essence of each species covered, as well as provide behavioral cues that may help the reader see the bird more clearly, or understand the bird's niche in the environment more fully.

It is our hope that this book will help introduce you to the marvelous variety of birds and habitats that grace coastal Northern California, one of the most diverse and enchanted environments on the North American continent.

ACKNOWLEDGMENTS

The late Rich Stallcup, friend and mentor, provided the inspiration and content of earlier drafts of this book. The manuscript benefited from careful reading and thoughtful commentary by two anonymous reviewers. The section on molt was reviewed and improved by Peter Pyle. The staff at University of California Press, especially Kate Hoffman, Kim Robinson, and Stacy Eisenstark, patiently shepherded the book through production, and David Peattie of BookMatters put on the finishing touches. Photographs were generously donated by Len Blumin, Seth Bunnell, Eric Horrath, Sean McAllister, Ian Tait, and Danika Tsao, and Keith Hansen provided his superb pencil drawings.

INTRODUCTION

This contribution to the California Natural History Guide Series of the University of California Press follows a long tradition of books that explain, explore, and celebrate the natural riches of California and beyond. Our intent is to tell beginning birders, or curious naturalists, the how, what, when, where, and why of birding.

Because birds are so mobile, some individuals of most species can wander far from their natal homes and appear anywhere. Here we have tried to include only those species most likely to be seen along the coast, from Big Sur to the Oregon border. This is not a field guide to bird identification, but a field guide to the birds themselves.

Birding is a word that encompasses many concepts. For some, the activity of searching for and observing birds is a clear window into the natural world, an affirmation of its beauty and its peacefulness. To others, birding is a delightful diversion from the hectic or perhaps boring daily routine of the modern world—providing calm amid the chaos. Birds are nature's ambassadors, connecting us through their ancient lineage to evolution's astonishing creativity and offering us some guidance, through our study of their habitat needs, in our stewardship of the Earth. Some people have found the wonders of birds to be the perfect antidote to sadness or loneliness, or a path to comradeship with kindred spirits; others consider the complexities of identification or behavior an intellectual challenge. To many of us, birding provides all these comforts and challenges.

Finding and identifying new species can be like treasure hunting, and the quest can be casual or all consuming. The treasure, once found, is a mixture of beauty, freedom, spirit, and a greater understanding of nature's sublime diversity. At the same time, even the most familiar birds offer opportunities for learning, discovery, and enchantment. Variations in the robin's song, the foraging techniques of egrets, and the flight patterns of swallows all display improvisation and provide the keen observer with a lifelong educational and aesthetic pursuit.

Getting started is easy. All that is needed is a pair of binoculars (preferably 8 or 10 power), a field guide, and the time and desire to go outside. A backyard, city park, or nearby wetland can provide days, months, or a lifetime of discovery and pleasure. Any place where nature abides hosts an ever-changing kaleidoscope of birdlife. Then again, you may also want to wander widely to visit national parks, preserves, or refuges, or travel to faraway lands to see exotic species.

Watching birds may take some practice, but the rudimentary skills are fairly easily acquired and available to almost everybody. It may be best to start with larger, slower-moving birds. It's easier to observe a Great Blue Heron stalking the edge of a marsh than a Chestnut-backed Chickadee flitting through the foliage. Once you are used to using binoculars and standing still for a few moments, birding techniques will come naturally. We humans like to categorize things, to pigeonhole them. Do not worry about trying to identify every bird you see, especially in the notoriously difficult groups like gulls or small sandpipers. Just admire their energy and watch them carefully with open eyes, and eventually, the distinctive movements and field marks of each species will reveal themselves to you.

This book covers birds that occur along the coastal strip of Northern California, with an emphasis on the most commonly occurring species. Although some coastal field guides consider only species found within a mile or two of shore, we extend coverage somewhat farther to include species observed on pelagic birding trips, especially to Monterey Bay, the Farallon Islands, and the Cordell Bank. Those unique destinations support such a diversity of species—marine birds that represent a large proportion of our regional avifauna—that to omit them would be an oversight. However, we emphasize the more common species and those more likely to be seen from shore.

Although you may think first of waterbirds when considering coastal birds, land birds are also important members of the coastal community. Vultures and ravens patrol the beaches for shore-cast carrion and human refuse. Some songbirds are restricted in distribution to the coastal fog belt or to coastal scrub, prairie, strand, and dune habitats. Peregrines and Merlins shadow the shores of estuaries in search of waterbird prey. Migrant land birds, especially young birds on their first migratory journey, follow coastal topography and "pile up" at islands of vegetation at headlands, lighthouses, and coastal promontories—a phenomenon known as the "coastal effect." Some coastal land uses, especially cattle ranching and other agricultural cultivation, have created habitats that attract large flocks of ground foragers—blackbirds, cowbirds, starlings, and even longspurs—that might not otherwise favor the coast.

The diversity of land birds is high in coastal Northern California, but many also occur inland, in drier environments. Within the discussion of each avian family, we have tried to include those land birds most characteristic of the coastal counties, or those species most likely to be encountered along the shore. Each discussion of avian families is followed by more in-depth species accounts. For these we selected especially representative coastal species (or subspecies) over those that are more wide-ranging geographically.

Birding basics are covered well in other books and not repeated here, but we expand some basic topics—for example, plumage and vocalizations—to provide more in-depth coverage. For the same reason, we discuss superspecies complexes, subspecies, and racial distinctions within and among birds unique to coastal Northern California.

Boundaries

The "Northern California" of our book title is a bit of a misnomer because we include much of what many call the "central coast." The geographical area covered in this book ranges from the southern border of Monterey County (35.8°N) northward to the Oregon border (42°N), nearly half the length of California's 840-mi coastline. The area includes two large estuaries—San Francisco Bay and Humboldt Bay. If their serpentine tidal contours are factored in, the area includes well over 1,000 mi of California's 3,427-mi tidal shoreline.

Many of the 13 counties that make up the region are larger than some states and, in terms of bird species, have greater biodiversity than most. The coastal counties considered here include, from south to north, Monterey, Santa Cruz, San Mateo, San Francisco, Alameda, Contra Costa, Solano, Napa, Marin, Sonoma, Mendocino, Humboldt, and Del Norte (see map on page 184). Those more interior counties (e.g., Napa, Solano) are included because their positions within San Francisco Bay subject them to tidal influence from the Pacific Ocean and they are directly exposed to the marine climate generated by the California Current. Considering the region as a whole, and counting the oceanic waters out to the continental shelf, well over 500 species of birds have been documented in the region.

These counties are in the Coast Range Bioregion (Evens and Tait 2005), a diverse mix of terrestrial and strictly coastal habitats. Some of the interior reaches of Monterey and Sonoma Counties, for example, are extremely xeric and support near desertlike conditions, as well as dry-country birds like Greater Roadrunner *(Geococcyx californicus)* and California Thrasher *(Toxostoma redivivum)*. But this is a book about coastal birds, and we will ignore the more interior reaches (no matter how interesting the birdlife) and concentrate on the regions of each county that are influenced by the coastal climate. If you are interested in more comprehensive coverage of California's rich avifauna, please refer to other books in the California Natural History Guide Series, especially No. 83, *Introduction to California Birdlife* by Evens and Tait (2005).

The coastline is quite well defined, but how far offshore does the coverage of this book extend? Many ocean birds are highly pelagic, rarely seen from land. Many of these are briefly noted in the species accounts, or referred to in the accounts of more thoroughly covered species. We tried to include those species that nest near shore, even if they are not likely to be seen by even the most intrepid observer. For example, several species of storm-petrel nest on islets and sea stacks within sight of headland overlooks, but these sea sprites arrive at and depart from their nesting cavities under the cover of darkness, a behavior thought to reduce the risk of predation by gulls and raptors. These nocturnal pelagic species can be very difficult to observe from land, but occasionally a strong storm or other anomalous weather event will drive the birds shoreward and afford good viewing for the fortunate naturalist. The authors have encountered such phenomena, though rarely, and have included some of those episodically occurring oceangoing birds in the species accounts (see Ashy Storm-Petrel, Red Phalarope).

Climate

The California Current dominates the climate of coastal Northern California. What controls the California Current? Massive atmospheric air masses that descend at the equator and rise at the temperate latitudes generate the North Pacific Gyre, a vast oceanic current that circulates clockwise around the North Pacific from the equator north to about 50°N, the latitude of Vancouver Island, British Columbia. The edge of this great gyre spins off smaller "boundary currents" along its outer edge. One of these, the California Current, is born in the Gulf of Alaska by the winds unfurling clockwise off the gyre. The gyre's winds drive the surface waters of the eastern Pacific in a southeasterly direction, parallel to the coastline of Northern California. As they travel south along our shores, these wind-driven surface waters are deflected offshore by the Earth's rotation—a phenomenon known as the Coriolis effect. As the surface waters shift offshore, they are replaced by colder waters from deeper in the ocean, a process known as upwelling. These cold upwelling waters are highly oxygenated and productive, supporting a rich community of plankton, the basis of the marine food pyramid. The cold waters of the current also cool and saturate the near-shore marine air and account for the persistent fog banks that shroud the coastline much of the year.

The California Current's cool marine air moderates the climate; summer temperatures tend not to be very hot, and winter temperatures not too cold (table 1). Rainfall increases from south to north, but precipitation is concentrated in winter months, especially November through February, at least historically. (Spring and early summer precipitation has apparently increased in recent decades.) The latitudinal increase in precipitation is expressed in the habitat types, with moist coniferous forests becoming more extensive to the north as annual rainfall increases.

The upwelling period dominates the coast's climate from March into August, accounting for relatively windy spring and cool summer temperatures. The persistence of the upwelling pattern, when cold water and increasing day length promote phytoplankton blooms, is of utmost importance to the ocean's productivity. As the Pacific High pressure system stabilizes in late summer, the California Current abates somewhat, upwelling decreases, and warmer surface waters from offshore move shoreward—a system called the "oceanic period," usually lasting from late August well into October. During these months, the north coast climate is most benign—storms are rare to

TABLE 1 Average annual precipitation and range of average annual temperatures from representative localities, south to north.

Site	County	Precipitation (in.)	Temperature (°F)
Monterey	Monterey	19.9	48.0-65.0
San Francisco	San Francisco	22.3	51.4-65.1
Mendocino	Mendocino	40.7	48.1-58.6
Arcata	Humboldt	41.9	46.4-55.4
Crescent City	Del Norte	66.8	44.7-59.5

Source: Data from the National Oceanic and Atmospheric Administration.

nonexistent, winds are negligible, and temperatures are mild. The equable autumn weather is followed by the Davidson Current period, from November through February, when warmer subsurface waters move northward between the California Current and the coast. The Davidson period corresponds to the timing of highest precipitation in each of the north coast counties. These various currents exert strong influences on the weather patterns as well as on the ocean's productivity.

It would be comforting to us all, human and bird alike, if the weather patterns were as predictable and regimented as the foregoing discussion suggests. However, the ocean's typical patterns are occasionally disrupted by weather anomalies of global proportions. A periodic weakening of the atmospheric pressure gradient between the Pacific Ocean and the Indian Ocean, known as the Southern Oscillation, seems to occur irregularly every several years and produces atypical ocean temperatures associated with those phenomena we know as El Niño and La Niña. These changing conditions have profound effects on sea surface temperatures (SSTs) and, in turn, on the productivity of the marine waters and the coastal climate. The warmer waters associated with El Niño may initiate a collapse of the marine food web and die-offs of marine birds. The warmer SSTs may also generate high-intensity winter storms, an additional threat to marine birds. La Niña's cooler-water episodes may have the opposite effect, with enhanced ocean productivity, but the effect on precipitation is more variable.

For all its potential variability, the climate of Northern California is relatively moderate. This mildness, coupled with its geographic position in the temperate zone and the diversity of habitats provided by the coastal topography, provide conditions that make the region a hotbed of birdlife.

Habitats

Ocean and Shoreline

The marine environment along the north coast, under the dominant influence of the California Current, is relatively stable compared with the terrestrial environment—temperatures vary within a relatively narrow range, and the upwelling tendency is very high, supporting a fairly reliable productivity base most years. The ocean's habitat can be partitioned into several zones, based on distance from the shoreline, proximity to the continental shelf, depth, and bathymetry. The terminology of coastal geomorphology is complex and varied (fig. 1), and not all taxonomies are in agreement.

Terms such as *inshore* and *nearshore* are generalities, at best, and there is no single definition for each, but they can be useful terms for describing the habitat preferences of birds and for describing marine habitats with somewhat vague boundaries.

Inshore waters, also called the littoral zone, are those closest to the shoreline and generally rather shallow, from the tideline to areas of the ocean beyond the breakers, but within sight of land, roughly a half mile under ideal viewing conditions. Typical birds of inshore waters are loons, grebes, pelicans, and scoters.

Nearshore waters are somewhat deeper and extend farther offshore and, for the purposes of this discussion, are the area influenced by longshore currents (the upwelling zone). In Northern California, the nearshore zone may extend to the edge of the continental shelf. Many seabirds that habituate nearshore waters may also be found inshore. Typical species include Common Murre, Sooty Shearwater, and Brandt's Cormorant.

Pelagic (from the Greek word meaning "open sea") waters, synonymous with *offshore*, are those waters from the continen-

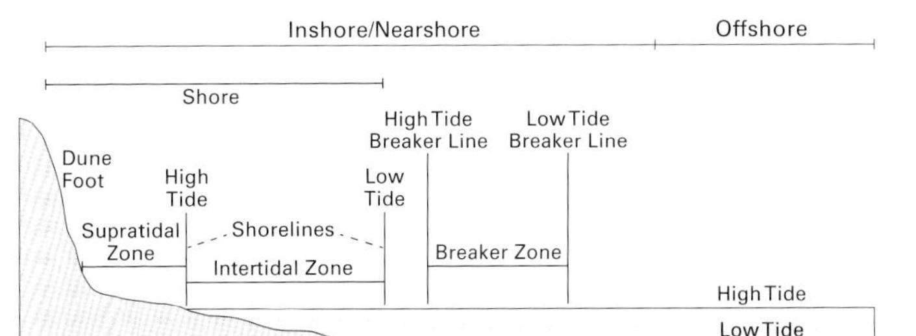

Figure 1

tal shelf and beyond, and pelagic birds are those that spend most of their lives on the open ocean. This is the realm of albatross, storm-petrels, and puffins, among others.

For this book, we will simplify the terminology and use *shoreline* to include supratidal and intertidal habitats, *nearshore* to refer to portions of the ocean within sight of the shore, and *offshore* (or *pelagic*) to signify waters not visible from the mainland.

Tides

The intertidal zone, or shoreline, is the boundary between land and sea. In contrast to the tides of the Atlantic (and much of the world)—which almost always occur twice daily with relatively little difference between successive highs and lows—the Pacific Ocean has a more complex tidal regime. The daily patterns of rising and falling water along the Northern California coast are called mixed tides (or semidiurnal mixed tides) and have relatively greater differences between successive highs and lows. Here, there are two high tides in a 24-hour period, one being higher than the other, and two low tides, one being lower than the other. This Pacific pattern, also known as "diurnal inequality," has produced the confusing terms *high-high tides* and *low-low tides*, best understood by studying an idealized graphic of a typical 24-hour tidal cycle (fig. 2).

For the curious naturalist visiting the coast, familiarizing yourself with the local tide is a critical consideration for ensur-

Tide Chart

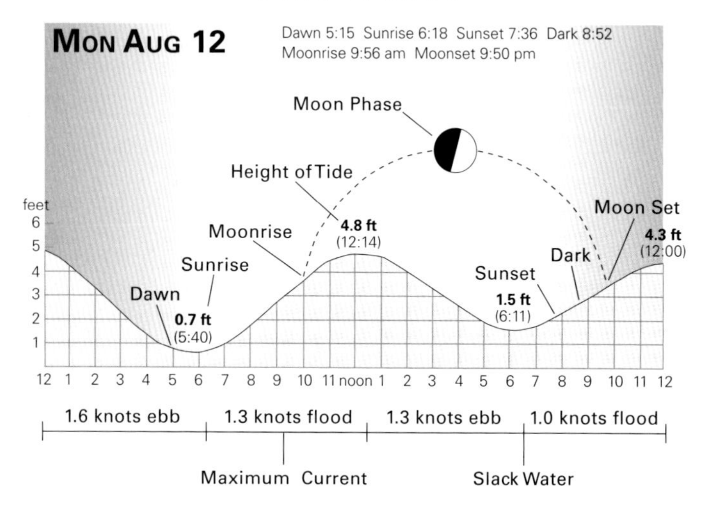

Figure 2

ing that coastal birds will be most closely observed, but also for your personal safety. When you go into the field to observe waterbirds, it is important that you consult a tide log and plan accordingly. Misjudging the tide can be fatal. Every year in Northern California, fishermen, beachgoers, and hikers get swept off the shore by "sneaker" waves or trapped on some offshore rocks by an incoming tide. In most estuaries—some of the premier birding spots discussed in Birding Opportunities and Roadside Nature Centers, later in this book, are tidal estuaries and lagoons—it is best to arrive when the tidal flats are exposed and the tide is incoming. Many shorebirds feed at the tide's edge, and as the tide rises, they are pushed closer and closer to shore, thereby affording the viewer more intimate looks. As the tide covers the flats, foraging shorebirds may retreat to high-tide roosts, such as adjacent salt marsh, levees, or islands, or shift to the outer beach, to either forage or roost. Waterfowl, on the other hand, need some depth of water to forage and raft, therefore as the tide comes in, the waterfowl too will be closer to shore and more closely observed.

MARINE SANCTUARIES

Generous proportions of our coastal waters are set off as National Marine Sanctuaries protecting one of the world's most diverse ecosystems and productive seabird nesting and foraging areas. The Monterey Bay National Marine Sanctuary (NMS) stretches from south of Monterey County to Marin County, encompassing a shoreline length of 276 mi and 6,094 sq mi of ocean. The Gulf of the Farallones NMS and the adjacent Cordell Bank NMS add another 1,811 sq mi combined to the north. (See fig. 3.)

The coastline includes a diverse mix of habitats—rocky shore, islets, and sea stacks, sweeping sandy beach, dunes and coastal swales, wide river mouths and coastal plains, tidal inlets and estuaries, coastal lagoons and embayments. The variety of habitats on the Northern California coast makes it one of the most attractive and biologically productive regions of the world.

Bays and Estuaries

Officially called the "closed waters zone," the bays and estuaries of the north coast are bodies of water where freshwater mixes with seawater to create some of the region's most productive and sensitive habitats—emergent wetlands, tidal flats, and eelgrass beds ("sea grass meadows"). Among the diverse mosaic of habitats on the Northern California coast, the estuaries, in particular, are important to birds. This reality is reflected in the official recognition of these estuaries by international organizations and the federal government. San Francisco Bay is considered a Site of Hemispheric Importance for Shorebirds, the highest possible ranking, by the Western Hemisphere Shorebird Reserve Network. Significant portions of San Francisco Bay and Humboldt Bay are incorporated into the National Wildlife Refuge system. Bolinas Lagoon and Tomales Bay are designated Wetlands of International Importance by the Ramsar Convention on Wetlands because of their value to wintering birds, among other criteria.

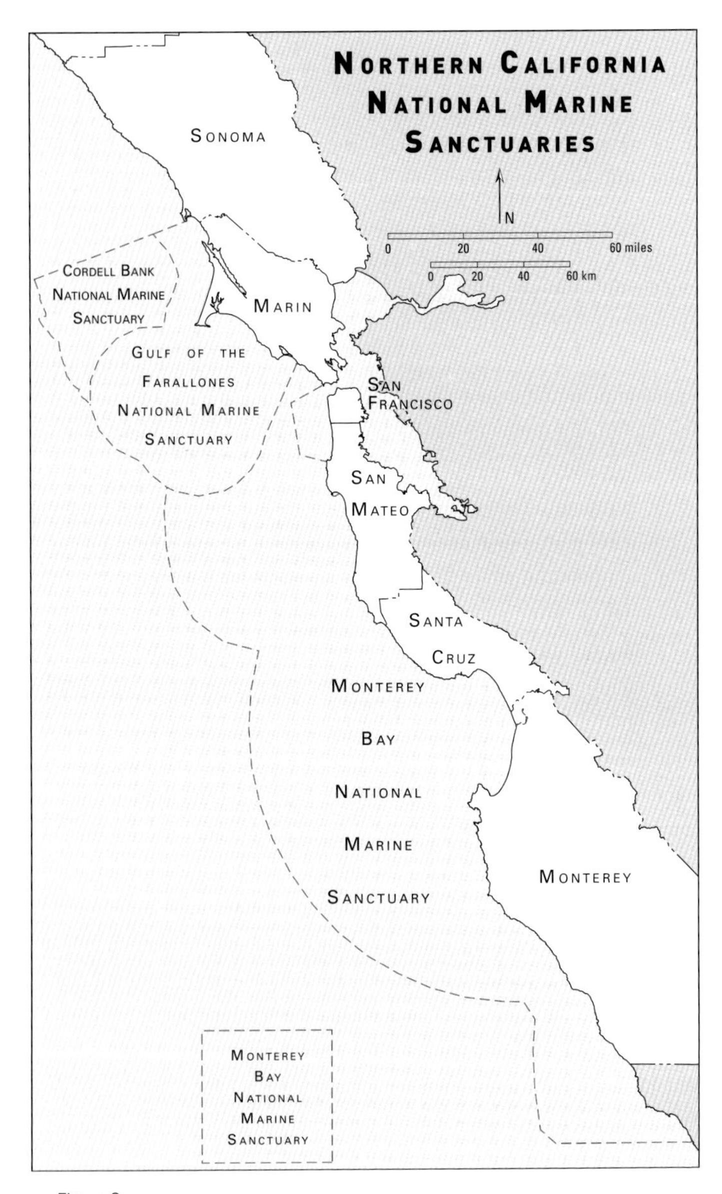
Northern California
National Marine
Sanctuaries
N
0
20
40
60 miles
0
20
40
60 km
Sonoma
Cordell Bank
National Marine
Sanctuary
Marin
Gulf of the
Farallones
National Marine
Sanctuary
San
Francisco
San
Mateo
Santa
Cruz
Monterey
Bay
National
Marine
Sanctuary
Monterey
Monterey
Bay
National
Marine
Sanctuary

Figure 3

Tidal Marshes

The vegetated intertidal zone and adjacent upland transition zones fringe the landward edges of the estuaries. Tidal marshes provide a productive ecological function for the estuarine environment, acting as filters for upland runoff, nursery grounds for a variety of fish and invertebrates, and foraging and nesting grounds for birds. The value of these habitats has increased as their extent has been reduced since the European colonization of California. The tidal marsh habitat of San Francisco Bay, by far the largest estuary in our region, was reduced by 78 to 88 percent between 1850 and 2000. In Humboldt Bay, an even larger proportion of historic tidal marsh was diked and drained for agricultural uses during European settlement. What were once tidally driven systems characterized by a salt-brackish-freshwater mosaic of marshland became highly fragmented with dikes and levees, isolated from tidal influence, and converted to uses for agriculture, industry (e.g., salt evaporation ponds), landfills, and urban infrastructure. Happily, since the last decades of the 20th century, some of these degraded historic marshlands have been (and are being) restored to tidal influence and regaining some of their former natural value. In San Francisco Bay there are numerous marsh restoration projects, with the ultimate goal of 100,000 acres of rejuvenated tidelands. In Humboldt Bay, the Arcata Marsh and Wildlife Sanctuary (307 acres), dedicated in 1981, serves as an object lesson in restoring the productivity to wetlands formerly covered by a landfill, lumber storage, and sewage oxidation ponds. At the south end of Tomales Bay, 550 acres of historic wetlands that had been diked for pastureland in the 1940s were returned to tidal action by the National Park Service in 2008; the site is rapidly reverting to natural habitat and a bird-rich environment.

Some characteristic birds of the tidal marsh are Song Sparrow, Marsh Wren, Common Yellowthroat, California Black Rail, and “California” Clapper Rail.

Tidal Flats

Tidal flats are nonvegetated, soft-sediment habitats in the shallow intertidal shoals of estuaries and other low-energy marine environments. Highly productive and nutrient rich, with dense concentrations of organisms living in the mud (infauna), tidal flats are primary foraging areas for myriad waders and other waterbirds and account for the large flocks of Arctic shorebirds that pass through during migratory periods, many of which remain through the winter months. Tidal flats also provide a buffer between deeper waters and the vegetated shoreline, dissipating wave energy and reducing erosion along the shore. Many of the shorebirds you will see foraging at the edge of the tide probe the mudflats for invertebrates, each species having developed it's particular bill length and shape, and feeding method, to exploit the various prey items that burrow in the substrate. Thus we see different-sized "probers," each sharing the same habitat and finding sustenance (pl. 1).

Not all tidal flat specialists are probers; those with larger eyes and relatively shorter bills, like the Black-bellied Plover (pl. 2), find prey by sight and pick it off the surface. Some waterfowl, most notably Green-wing Teal, siphon food off the damp surface of the mud, and egrets and herons poke around shallow standing pools or the vegetated edge of the tidal flat for whatever morsels they can find.

Salt Pannes and Salt Ponds

San Francisco Bay holds the largest extent of this habitat type in our region. Before European colonization, the bay had about 8,000 acres of natural salt pannes—large, shallow saline ponds scattered along the backshore of tidal marshes. Salt pannes were, and are, particularly attractive habitat for long-legged waders—avocets, stilts, and yellowlegs. With urbanization of the estuary and an increase in the demand for salt in the mid-1800s, salt ponds were constructed, replacing tidal marsh and associated habitats, in effect mimicking the native salt pannes. By the beginning of the 21st century, salt ponds covered nearly 35,000 acres. These salt ponds, and the associated levees, are attractive to large numbers of shorebirds as foraging areas and to several species as nesting habitat—most notably American Avocet (pl. 3); Black-necked Stilt; Western Snowy Plover; California Gull; and Forster's, Elegant, and Least Terns.

Salt ponds vary greatly in their value as bird habitat, depending on differences in salinity, size, depth, and the resulting community of invertebrates. Nevertheless, in their entirety, shallow saline ponds and those pannes that still exist are bird-wealthy environments. Undoubtedly the abundance of locally nesting shorebirds—especially American Avocet and Black-necked Stilt—has increased with the expansion of salt-pond habitat. Western Sandpiper, too, once described as a "sparing winter visitor" (Grinnell et al. 1918) in San Francisco Bay, has become the most abundant wintering shorebird in the Bay Area, concurrent with the development of artificial salt ponds.

Open Bay Waters

Protected from Pacific swells, open bay waters provide rafts of waterbirds with roosting areas and foraging opportunities. In the larger embayments, large flocks of canvasbacks, scoters, scaup, buffleheads, grebes, coots, and gulls are often found loafing in leeward waters just offshore. These waterbirds gather in October and remain through the winter months, sometimes forming vast flocks, especially in protected coastal bays where hunting is not allowed. San Francisco, Humboldt, and Tomales Bays support large proportions of populations of some species during winter, and provide safe refuges (in some portions) from depredations by hunters in fall.

In each of the larger bays, submerged pastures of eelgrass *(Zostera marina)* provide critical foraging habitat for Black Brant, a sea goose, but also for diving ducks—especially scaup and scoters—that forage on herring roe and other invertebrate prey that abounds within the eelgrass beds.

Outer Coast: Beaches and Rocky Shore

Although the sandy beaches may appear to be barren, if beautiful, habitat, they are actually dynamic ecological engines. Walk the beach in the morning after a high tide and notice the wrack line, tangled debris of marine algae ("seaweed"), shore-cast carrion, and discarded crab carapaces. Often the tracks in the sand show that foxes, coyotes, skunks, raccoon, gulls, ravens, or vultures have already investigated the newly arrived bounty for edible morsels. Lift a blade of bull kelp and note the teaming amphipods and beach hoppers, already busy decomposing the detritus. And this is just the life visible to the naked eye.

Sandy beaches support a somewhat more modest community of shorebirds than the tidal flats of the estuaries. The emblematic bird of the outer beach is the Sanderling (pl. 4), a chunky, pot-bellied, Arctic-nesting shorebird that spends the nonnesting season along our shores, chasing the receding waves out to forage in the swash zone, then racing back up the beach before the next wave washes in. Willets, too, are common shorebirds of the outer beach, larger than the Sanderlings and more aggressive, often trying to steal morsels from their smaller cousins or from others of their own kind. The main prey for these sandy beach shorebirds are mole crabs, also called sand crabs *(Emerita analoga)*. Like Sanderlings, mole crabs occur on most beaches in North and South America. No larger than a child's thumb, these burrowing crustaceans migrate up and down the beach in the zone of wet sand washed by waves. Unlike most crabs, these have no claws. Efficient burrowers, they stay mostly buried in the damp sand, always facing seaward, only their eyes and antennae protruding. As the waves wash in, they unfurl their long antennae and filter plankton from the seawater. Mole crabs can be incredibly abundant, with several thousand individuals in every square yard of beach, which explains the abundance of Sanderlings.

The outer beaches and associated foredunes are also critical habitat for the federally threatened Western Snowy Plover, a small, sand-colored shorebird that nests, though sparsely, along the more remote sections of our beaches. Increasing coastal development has eliminated or greatly reduced plovers from many of the north coast beaches (Pajaro Dunes in Monterey, Stinson Beach in Marin). Small numbers of plovers persist, and a few of those are still able to fledge young on beaches actively protected by public agencies—for example, Point Reyes National Seashore (Marin Co.) and MacKerricher State Park (Mendocino Co.)—if they are able to evade the depredations of the marauding Common Ravens, whose numbers have increased dramatically in recent decades.

Most rocky shore habitat is on the rugged outermost coast, at the foot of steep headlands, pinnacles, pillars, tombolos, islands, islets, and sea stacks, confronting the brunt of the Pacific's most powerful elements.

Compared with the sandy beach, the rocky shore is a harsh environment, continually pummeled by waves and crowded with dense communities of tough invertebrates—mussels, limpets, barnacles, and sea stars—whose entire evolution has focused on holding fast to the rocks. Most of the birds that frequent this habitat, many habitually and exclusively, have evolved with equal prowess to pry, chip, or chisel these tenacious animals loose. The physical model for each avian "rock star" is similar: stout bill, thick legs, relatively heavy body, and large eyes, all the better to see you with in a misty, low-light environment. Most of these rocky shore birds sport dark gray or black plumage, blending in with the generally dark pallet of the rocks on which they roost and forage. Perhaps the most emblematic bird of this habitat, and the only shorebird that nests in this extreme environment, is the Black Oystercatcher (pl. 5).

The methods used for extracting meat from their tough prey items are as varied as the predators' bills: oystercatchers jab mussels and sever the abductor mussel or use the bill as a lever to pry chitons loose; turnstones peck and probe beneath algae or in crevices; Surfbirds tug, pull, and peck, breaking barnacles loose and swallowing them whole. Whatever works.

Nearshore Waters

For the purposes of this book, the "nearshore waters" include the open ocean from the low-tide line out to the visible horizon. Of course this distance varies with weather, the elevation of the observer, and the power of the optics being used, but it's a convenient concept. A more formalized approach identifies the nearshore zone as marine waters up to 328 ft in depth, so there is a large overlap between the formal definition and this approach. The nearshore waters in Northern California are relatively shallow and overlie the continental shelf, the gently sloping margin of the continent that extends offshore until it breaks and drops off precipitously at the continental slope. The shelf is relatively narrow in most sections of our region but varies from 4 to 20 mi in width. (Worldwide, continental shelves average about 40 mi in width.) But the topography of the ocean floor is as varied as that found on the land; in a few places, submarine canyons, with depths up to several thousand feet, come closer to shore. Monterey Bay is the most prominent example, where the deep water of Monterey Submarine Canyon (fig. 4) bisects the continental shelf and approaches within a few hundred feet of shore.

The dominant habitat types of the nearshore waters are

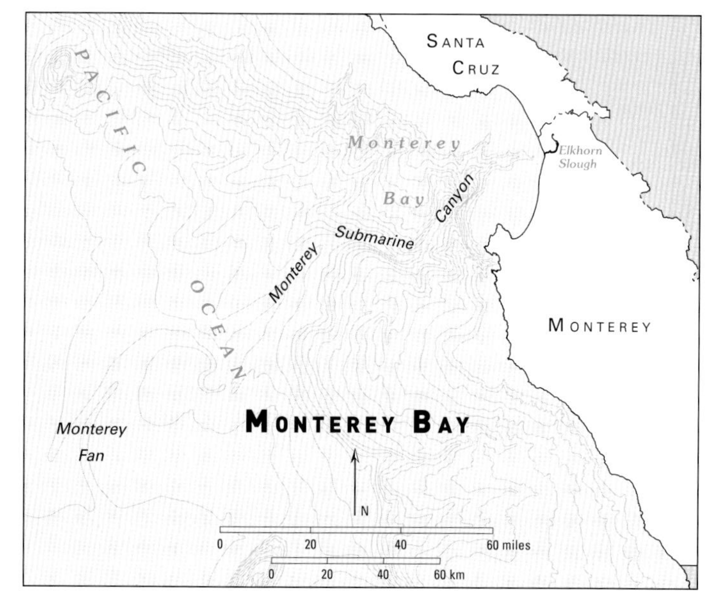

Figure 4

submerged boulder fields; seafloor composed of shale, sand, or mud; and kelp forests. Boulder fields occur off headlands and resistant geological formations where the wave energy is most intense. Sandy bottoms occur intermittently along the coast, with finer sands accumulating in shallower waters where wave action is gentle, and coarser sands accumulating where wave energy is more intense. Muddy bottoms occur where silt and clay settle out from land runoff, at the mouths of rivers and estuaries. (Also, the percentage of mud, silt, and clay increases as depth increases on the shelf.) Kelp forests are common in areas with rocky substrates, whether off headlands or along calmer shores, and may extend for miles along the coast in waters up to about 100 ft deep. Marine birds are often associated with these different habitats as determined by the nature of the bottom. Pelagic Cormorants, for example, search for small octopi amid the crevices of boulder fields. Harlequin Ducks, too, seem to prefer rocky shorelines and substrates with boulders and cob-

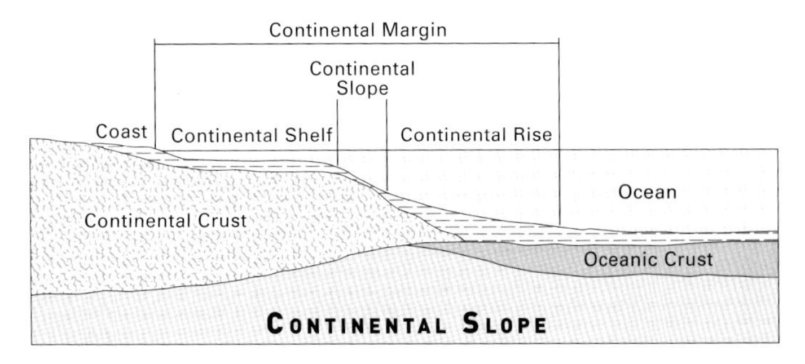

Figure 5

bles. Common Loons tend to favor calmer water with sandy or muddy bottoms for pursuing schooling fish. Great Blue Herons may use the floating bulbs of kelp as perches for hunting Kelp Pipefish *(Syngnathus californicus)* or even a Sarcastic Fringehead *(Neoclinus blanchardi)*. Surf Scoters dive among the stalks for sessile crustaceans, and turnstones search the shore near kelp beds, plying the windblown blades for amphipods.

Farther out, but within sight of land, the larger, flocking fish eaters (Brandt's Cormorants, Brown Pelicans, and Western and Clark's Grebes), and of course the opportunistic gulls, have a common nearshore presence. The nearshore zone is also the bailiwick of the coastal nesting seabirds—Common Murres, Pigeon Guillemots, Marbled Murrelets.

Offshore Waters

This oceanic zone begins beyond sight of land at depths of 300 ft or more and extends beyond the continental shelf. To reliably experience its birdlife, as well as the marine mammals that occur here, a boat trip out of one of the coastal ports is recommended—from Monterey Bay, San Francisco, Bodega Bay, or Fort Bragg. The "featureless plain" characterization of the Pacific Ocean is an illusion. Birds and other marine life tend to concentrate where upwelling is strongest or currents converge, in the vicinity of submarine canyons or submerged banks, or where the shelf breaks at the continental slope (fig. 5). The submarine topography causes mixing of waters and nutrients that generate productivity of microscopic organisms (phytoplankton and zooplankton) that form the basis of the oceanic food chain.

Taxonomy: Subspecies, Species, and Superspecies

Taxonomy is the science of classification of animals and other organisms sorted systematically into categories that ladder down from the most general to the most specific (from the same Latin root as the word *species*). For example, a Brown Pelican (pl. 6) is in the:

Kingdom Animalia
Phylum Chordata
Class Aves
Order Pelicaniformes
Family Pelecanidae
Genus *Pelicanus*
Species *occidentalis*

A species is the ultimate expression of evolution. Speciation does not happen suddenly, but unfolds through the dynamic interaction of groups of organisms with the environment and the selective pressures exerted upon them. To explain the variation within and among species, taxonomists have come up with many concepts, two of which—*subspecies* and *superspecies*—are useful for framing the core concept of species and for understanding variation in birds. It is important to remember, however, that these classifications are somewhat speculative, often controversial, and always open to revision. Determinations about whether two populations should be considered different species or just different subspecies can be very difficult to make in some cases, even for taxonomists. (In fact, there are two dominant approaches to vertebrate taxonomy: the *biological species concept* and the *phylogenetic species concept*. Traditionally, birds have been evaluated by the former. See the glossary for definitions of each.)

Superspecies include pairs or groups of species that are very similar physically (phenotypically) but are geographically isolated from one another (allopatric) during the reproductive phase of their life cycle. Superspecies do not hybridize to any significant extent and have diverged from one another fairly recently in evolutionary time. When a superspecies includes two species (e.g., Common and Yellow-billed Loons), those species are called sister species or sibling species. When a superspecies group consists of several species (e.g., the pink-legged Pacific gulls), it is usually called a superspecies complex.

These are some of the superspecies that occur on the Northern California coast:

Common and Yellow-billed Loons
Western and Clark's Grebes
Tundra and Trumpeter Swans
Blue-winged and Cinnamon Teal
Greater and Lesser Scaup
Short-billed and Long-billed Dowitchers
California, Herring, Western, and Glaucous-winged Gulls
Spotted and Barred Owls
Rufous and Allen's Hummingbirds
Townsend's and Hermit Warblers

Subspecies, on the other hand, are groups of *phenotypically variable* (slightly different in plumage or measurements) members of the same species, with each of the subspecies essentially equivalent to a geographical race. A subspecies is considered a "species in the making" because it represents a geographic segment of a species' population that has distinct differences in coloration, and perhaps size and physiology. These differences, in turn, often correspond to differences in habitat selection and behavior. Where the ranges of subspecies overlap, interbreeding will occur. Perhaps the most common example in our region is the Yellow-rumped Warbler, composed of two subspecies, the Myrtle and the Audubon's. Another example is the Song Sparrow of San Francisco Bay: three distinct subspecies—Alameda, San Pablo, and Suisun—are resident in different geographical areas of the salt marshes around the bay shore.

So, superspecies and subspecies bracket the principal and most basic category of taxonomic and biological classification, the species. (The word *species* is both singular and plural.) A species is recognized based on its geographic distribution and its physical characteristics—where it breeds and what it looks like. The determination and naming of species status is an ongoing scientific endeavor, reflecting the changing nature of our knowledge and the changing nature of the environment. As genetic analysis becomes more sophisticated and precise, new species are discovered within a given genus and a single species is "split" into two, or sometimes more. Genetics (i.e., DNA sequence analysis) has become a primary tool in taxonomy; however, natural history—especially breeding behavior, vocalizations, and habitat preferences—also plays an informative role in the assignment of species (and subspecies) status. For North American birds, the final arbiter of taxonomy is the American Ornithologists' Union (AOU) Committee on Classification and Nomenclature, a group of professional ornithologists that reviews the literature and the evidence and is continually revising the list of North American birds. The taxonomy used in this book follows the seventh edition of the AOU Check-list of North American Birds through the 53rd supplement (2012). (The AOU has relied mostly on the biological species concept in categorizing North American birds.)

Each species is given a two-part name composed of a generic name and a specific name. Thus, a Canada Goose *(Branta canadensis)* (pl. 7) belongs to the genus *Branta,* which includes several species, and the species *canadensis,* which identifies only one species. But, there are several subspecies of the Canada Goose, and each is given a third name to indicate it's subspecific status. The most common one in our region, the large "honker," is assigned the subspecies name *moffitti*; thus the full name is *Branta canadensis moffitti.* Currently, seven subspecies of Canada Goose are recognized, collectively known as "the greater Canada Goose complex." In 2004 the AOU split *Branta canadensis* into two separate species, the larger Canada Goose *(B. canadensis)* and the smaller Cackling Goose *(B. hutchinsii).* The way these species are divided is illustrative of the methods used to categorize birds by taxonomists. The Canada Goose includes larger-bodied, more southerly breeding populations, whereas the Cackling Goose is smaller bodied and breeds farther north, in the Arctic tundra. Also, notice the name of the new species: "cackling" describes one characteristic in which it differs from the larger group, the "honkers," illustrating how natural history (in this case vocalizations) as well as genetics inform decisions about taxonomic classification and nomenclature. Five subspecies are considered under Cackling Goose, sometimes referred to as "the lesser Canada Goose complex."

Not all species are as variable as the "white-cheeked goose complex" described above. Some species are ecologically or behaviorally isolated enough to maintain a discrete population. The Common Loon (pl. 8), for example, is considered monotypic; that is, there are no subspecies. Although there is considerable variation in size and the measurements of other characters (bill length, wing chord, etc.) within the species when it is distributed over an area with differing environmental conditions, variation tends to follow a gradual geographical distribution. That continuum of change is termed "clinal" variation. The observer would not be able to differentiate a Common Loon that breeds in northern Alaska from one that breeds in Oregon, though the more northern birds are apt to have slightly longer wings or bills than the more southerly nesting individuals.

Even monotypic species like the Common Loon can interbreed with closely related species in certain, and usually rare, circumstances. There are five species of loons, all in the same genus *Gavia*, thus they are called "congeners."

The Common Loon is most closely related to the Yellow-billed Loon, and as discussed above, the pair forms a superspecies. Indeed, hybrids have been reported between these two, as well as other, congeners.

These examples are given not to confuse the reader, but to illustrate that species boundaries are "somewhat mutable," that evolution is always testing biological fitness, improvising, searching for new shapes and sizes to adapt to a potentially changing environment. Evolution is a dynamic ongoing process.

Plumage of Birds

All birds have feathers, and they are the only animals that do. Simply put, feathers define birds. Feathers are an ingenious evolutionary invention that enables flight and provides insulation, waterproofing, and camouflage (cryptic coloration) as well as brilliant breeding plumage. Composed of the protein

FEATHERY TERMS

Down Soft and fluffy, without interlocking barbules, down provides warmth and buoyancy. Newborn chicks are covered with *natal down*. In adult birds, *body down* underlies the contour feathers of the plumage. *Powder down* is an uncommon feather type that grows continuously and therefore is not molted but sloughs off "feather dust" that serves to protect the other contour feathers. Powder down is found on members of the heron family, as well as on some parrots and pigeons.

Bristles Hairlike and stiff, bristles often surround the beak or eyes (as whiskers or eyelashes do in mammals); their function is not well understood but is likely sensory.

Filoplumes Also hairlike and occurring mostly on the nape or crown, filoplumes are thought to provide insulation as well as to have decorative and sensory functions.

Semiplumes Midway between down feathers and contour feathers, semiplumes add warmth and fill out the smoothness of the plumage.

Remiges The primary flight feathers of the wings, remiges are long, stiff, and attached directly to the bones with ligament. Note that on many larger birds—swans, geese, cranes, pelicans, gulls—the tips of the remiges are darker, with concentrations of melanin for added strength and durability.

Rectrices The tail feathers, rectrices are also stiff and strong, but only the central ones are attached to the bone by ligament. The rectrices are the rudders of the birds in flight, and they aid in balance and maneuverability.

beta-keratin, feathers have evolved into a variety of structures—from the flimsy topknot of a California Quail to the stiff wing feather of a Brown Pelican. There are two basic types of feathers—down and contour feathers—and each of these is modified into various types to serve specific functions in a bird's life.

Because waterbirds are a primary consideration of this book, a brief discussion of the uropygial gland is worthwhile. Also called the "preen gland" or "oil gland," it is located near the base of the tail (on the rump) and is present in most species but best developed in aquatic birds, such as petrels, pelicans, ducks, and the Osprey. The gland secretes oil that the bird applies to its feathers with its bill. This oil is a diester wax called uropygiol, hence the name of the gland. Although the wax helps protect the feathers—and perhaps the bill and legs—the act of preening also aids in waterproofing by creating an electrostatic charge to the feathers. Most waterbirds have well-developed oil glands, but the structure of their feathers also contributes to their ability to survive in their watery world. (Interestingly, the gland may play a role in sexual selection, because the secretion in females has been found to emit a specific odor.)

Molt (or Moult)

Feathers are fragile and need to be replaced regularly if they are to maintain their usefulness. Feather wear (abrasion), and therefore the process of feather replacement, is determined by a bird's life activity, which is determined in turn by its behavior, distribution, habitat preferences, and genetic heritage. The causes of feather wear are many: damage by contact with vegetation, rocks, and other birds; friction with the air during flight; and perhaps most important, loss of structural integrity through exposure to sunlight. In general, the feathers of birds that migrate wear more readily than those of nonmigratory (sedentary) birds; this may be more the result of exposure to solar radiation than of abrasion. Birds that live in habitats with intense solar radiation or harsh vegetation will experience more wear than those living in shady forests.

Molt is a cyclical process of feather restoration that all birds must undergo. During normal molt, a new feather begins growing within the follicle and pushes out the old feather. Thus, when a feather drops, it is already being replaced, leaving no significant gaps in the plumage. The main flight feathers, remiges and rectrices, are molted symmetrically on the right and left sides, so direct flight is balanced. (Some species, such as waterfowl, loons, and rails, molt all flight feathers at once and go through a brief flightless phase, however.) Although feathers grow through follicles as human hair does, the process is not continuous. Molting patterns are highly variable and have been organized into four basic "strategies." This is a complex subject, well beyond the scope of this book, but for the curious naturalist, an exciting area of exploration. Full treatment can be found in Pyle (1997, 2008) and Howell (2010).

Molt can be "complete," that is, all feathers are replaced, or "partial," meaning only certain feathers are replaced. Molting is energy intensive, and most species do not molt during those periods when their energy is needed for critical life functions—migration and reproduction. Most birds undergo complete molt after the nesting season, in late summer and early fall, but molt can be protracted, lasting well into winter and spring.

Once the Common Loon (pl. 9) completes its partial spring body molt, which will take several weeks, it will be in full breeding (or alternate) plumage. During the winter, when it is most likely to be seen in Northern California, it will be in a duller wintering (basic) plumage. The nomenclature of molts and plumages of birds is complex (some would say arcane) and undergoing revision as molting strategies are better understood. For the amateur birder, and for the purposes of this book, we will simply use the terms *nonbreeding* (or sometimes *wintering*) and *breeding* plumage to describe the annual feather and plumage cycles of birds. The more formal names for these feather cycles are, typically, *basic* (nonbreeding) plumage and *alternate* (breeding) plumage, regardless of the season during which the plumage is acquired. However, this general terminology should be reversed when considering the ducks (family Anatinae):

> The bright plumages of adult male ducks (other than Ruddy Duck) in late autumn and winter should be considered basic plumages, which are completely renewed annually as in other birds. The ephemeral and highly variable cryptic plumages found in spring and summer presumably have evolved more recently, primarily in those species benefiting from an ensuing camouflaged plumage, and thus should be considered the alternate plumages. (Pyle 2005)

It is also helpful to determine the ages of birds as revealed by their plumage. Juvenal plumage is the first covering of feathers that replaces the downy feathers of the chick. It is distinctly different from either alternate or basic plumage (pl. 76), and some groups of birds, particularly sandpipers, sport the juvenal plumage through the first summer and fall. In Great Britain, the term *juvenile plumage* is used; for all intents and purposes, the terms are synonymous. Here, the term *juvenal plumage* is used to describe *juvenile* birds that have not yet attained an adult plumage. It is part of the "first plumage cycle," which also includes "formative" (postjuvenile) and sometimes first-alternate plumages, considered "immature" here. Once a bird has fully matured, it will be referred to as an adult having "definitive" adult plumage or as being in the "definitive plumage cycle." Some species (e.g., gulls) exhibit identifiable second-, third-, or fourth-cycle plumages that we refer to as "subadult." Be aware that terms within these cycles continue to be a field of exploration (and debate) that awaits investigation by the curious naturalist.

Dimorphism and Monomorphism

Dimorphism literally means "two forms," and *monomorphism* means "one form." Sexual dimorphism, in which the male and the female of a given species differ in looks, is a common trait among birds. Among California's coastal birds, the waterfowl provide perhaps the most obvious examples of sexual dimorphism. The definitive adult plumages of a male and female Mallard, for example, are quite distinct and easily recognized. But dimorphism can be subtle as well. The sexes of many of the shorebirds differ in bill shape or size, but are virtually identical in plumage. In the Long-billed Curlew (pl. 11), the female's bill is noticeably longer than the male's bill, and each is shaped somewhat differently. The male's bill follows an even curve for its entire length, whereas a female's bill is somewhat straighter along its length but has a more abrupt curve at the end.

Perhaps no species are entirely monomorphic, but the term is used to describe those in which males and females are virtually indistinguishable, at least to human eyes. Adult Common Murres (pl. 12) show no discernable dimorphism, but the bill of the male is slightly longer and deeper than that of the female, and the wing may be slightly longer; however, neither character is obvious in the field. Most species that display little sexual dimorphism share a more egalitarian lifestyle; they tend to behave similarly and participate equally when nesting and chick rearing.

Seasons and Migration

Although the coast of Northern California is battered by winter storms, assailed by occasional gale-force winds in spring, and often blanketed with fog in summer, millions of birds of many species live here year-round. They are *permanent residents*, and during normal weather, survival and reproductive success is very high.

Many more birds of other species are *migrants*. Some that nest here fly to the American tropics or beyond during our winter, and others nest far to the north and east but come to spend the winter here. Other cohorts of birds (especially some of the plovers and sandpipers) are found on the coast during spring and fall only—just passing through.

Though we speak of four seasons, the timing of occurrence and passage of the many species varies and overlaps so much that it is impossible to define bird seasons by date. We have tried to provide the seasons of occurrence for each species in the Occurrence Charts at the end of the book. Birding along the Northern California coast can be exciting and productive any time of year, and any given month will contain its own community of birds to behold. That said, there are some generalities to keep in mind.

Winter

During winter, generally November through March and into April, coastal bays, harbors, lagoons, and river mouths may be crowded with ducks, grebes, and loons as herons and egrets patiently ply the shore and kingfishers watch the still waters from overhanging perches. While Sanderlings, Willets, Marbled Godwits, plovers, and flocks of gulls frequent beaches that are safe from disturbance, a Peregrine Falcon or Merlin is probably perusing the panorama from a distant perch.

Spring

During spring, generally March into early May, most wintering waterbirds depart and multitudes of migrant waterbirds pass through. Although the numbers of migrant waterbirds are countless, gross estimates are possible. Over a million Pacific Loons and perhaps one-third of a million Black Brant fly north over nearshore waters in April and May. Well over a million shorebirds—mostly Western Sandpipers, Dunlin, and dowitchers—move through the estuaries en masse during the last 10 days of April. Most of these marvelous birds, heading to the far north to nest, are dressed in their finest breeding plumages.

Summer

Summer, generally May and June, is the slow season for birds on the immediate coast. Although no bird season conforms to our Roman calendar, the bird summer is the most difficult to define. Anna's Hummingbird may begin its nesting season by early December and hatch chicks by the New Year. Clapper Rails set up nesting territories as early as February. Different species in different habitats begin and conclude reproductive efforts right on through the summer months, ending with late fledglings in October. So, the nesting seasons of our local breeders overlap with the migratory seasons of our winter residents and migratory visitors.

The first southbound migrant shorebirds appear in the last few days of June, adults having completed nesting efforts and still wearing their breeding (alternate) plumage, which they retain into mid-July, when the first juvenile shorebirds begin to arrive. These early migratory pulses are really the beginning of the fall season, at least waterbird-wise.

Fall

Fall is the most exciting bird season along the coast. Along with the astonishing numbers of winter arrivals and passing travelers, the possibility of encountering rare species is far greater than in any other season. At least half of the many millions of birds in motion are youngsters, "birds of the year," only a few months old. A small percentage of these are equipped with genetically faulty compasses that cause the birds, on their maiden voyage, to wander off course or follow the wrong bearing and end up outside the species' usual migratory pathways, far from where they are supposed to be. Migrating birds are subject to the "coastal effect," a tendency to pile up while following the coast, especially during fall migration. (Most land birds try to avoid crossing the air mass that overlies vast expanses of water.) This phenomenon is most pronounced in young birds, those embarking on their first long-distance journey, and adds a special level of excitement to the autumn migratory period.

Perhaps the most emblematic avian event of fall along our shores is the arrival of "the three amigos"—Brown Pelicans, Heermann's Gulls, and Elegant Terns—moving north from their nesting grounds in Mexico. They come to forage in the cool, nutrient-rich waters here, returning south, many to Mexico, by October.

Ethics of Birding

Life is not particularly easy for birds or other wild animals, and they generally perceive humans as predators, for good reason. Their lives are devoted to survival in an environment that is often hostile, and all of their waking energy is devoted to "making a living" and reproducing. When you observe animals, it is important to keep this reality in mind and to remember that "observation is not interaction." In other words, keep a respectful distance so you do not flush a bird or otherwise alter its behavior.

It is understandable that we may want a little better look, or a closer photograph, but if we cause the creature to fly, or scurry for cover, we are approaching too closely. We now have access to high-powered optics—binoculars, telescopes, and telephoto lenses—that allow us to get "up close and personal" with wild animals from a comfortable and respectful distance. If you can manage to spend the extra money for clearer or more high-end optics, you will not regret it, and you will reduce your need to intrude into a bird's personal space and increase the quality of your viewing experience.

Birds are most vulnerable to disturbance during the nesting phase of their life cycle. Approaching a nest, or removing vegetation that helps hide a nest, exposes the bird and its young to predation and nesting failure. Predators—especially jays, ravens, gulls, and foxes—cue on human scent and movement and may investigate human trails in search of food.

The famous axiom "leave only footprints" is also a good guideline to keep in mind when visiting natural habitats. In parks, preserves, or wildlife refuges open to the public, stay on the maintained paths and follow the rules. They are developed for a reason—to protect the natural resources and the wildlife from undue disturbance.

Respect private property and the privacy of residents. If you want to access private lands, make an effort to contact the landowners beforehand; explain your interest and why their properties may support interesting birdlife. Many people are happy to share their environment if approached respectfully. If access is refused, take "no" for an answer.

Wear muted colors and walk softly through the forest, around the marsh, or across the field. You will be more likely to go unnoticed, and will likely see more wildlife. Also, studies have found that wildlife in general (birds in particular) are less likely to flush when a human approaches at an oblique angle rather than straight toward the subject. So, when walking from your car to the shoreline of a lagoon to view a flock of shorebirds, do not walk directly toward them; rather, veer off to the side. You may also want to position yourself with the sun at your back, for better lighting and to be less visible to the birds you are observing.

Humans are social animals and often enjoy the company of others, but this natural inclination can disturb wildlife and cause animals to avoid an area where there is a high level of human activity. Best to travel in small groups, or even singly. The larger the group, the less you will see and the more impact you will have on the environment. If you are in a group, even a small one, speak softly, or not at all. Some of our most memorable birding experiences have occurred solo—sitting quietly by a pond at dusk, scanning a marsh while backlit by the morning sun, or having a glass of wine on the deck and watching the feeder and birdbath.

Which brings us to an oft-asked question: Is it OK to feed the birds? There are various views on this subject. Perhaps the best answer is to cultivate native food plants in your yard,

CITIZEN SCIENCE

Birding, or bird-watching, is a pastime, a sport, and an avocation that often develops into a passion. The skills of observation acquired through the pursuit of birds dovetail nicely with the scientific method, and the participation of "citizen scientists" in data collection enhances various scientific investigations.

Birders contribute to ongoing research projects that are nationwide in scope. National Audubon's Christmas Bird Count has been collecting information on winter bird populations throughout North America (and beyond) for well over a century. The North American Breeding Bird Surveys (U.S. Geological Survey and partners) monitor the status and trends of nesting bird populations. Observers who volunteer for Cornell Lab of Ornithology's Project Feeder Watch or the Great Backyard Bird Count provide indices of overwintering species.

There are also many regionally focused projects that involve citizen scientists, most of which are fostered by research or land management organizations. Most programs require some degree of training, but the hosting organizations are pleased to accommodate enthusiastic volunteers. Following are a few examples of ongoing projects that welcome citizen scientists:

Coastal California Shorebird Survey—San Francisco Bay, a cooperative effort between Point Blue Conservation Science (formerly PRBO), the San Francisco Bay Bird Observatory (SFBBO), U.S. Geological Survey (USGS), and Audubon California, documents numbers and distribution of wintering shorebirds across the extensive Bay Area wetlands. ▶

Save Our Shorebirds Surveys—Mendocino is hosted by the Mendocino Coast Audubon Society conducting seasonal censuses along that coastline.

Colonial Waterbird Survey (PRBO, SFFBO, and Audubon Canyon Ranch) monitors population trends of a variety of colonial nesting birds with a focus on herons and egrets in the greater San Francisco Bay Area.

The Landbird Project at SFBBO's Coyote Creek Field Station (Alameda County) enlists volunteers to study dispersal, migration, behavior, social structure, life span, survival rate, reproductive success, and population growth.

Hawk Watch (Golden Gate Raptor Observatory) monitors migrating raptors every fall, August to November, at Marin Headlands above the Golden Gate.

COASST is a citizen science project dedicated to involving volunteers in the collection of high-quality data on the status of coastal beaches and the trends of seabirds. COASST volunteers systematically count and identify bird carcasses that wash ashore along ocean beaches from Northern California to Alaska. Volunteers need no experience with birds, just a commitment to survey a specific beach (about 0.75 mile) each month.

Peregrine Falcon Nest Survey, a project of the Santa Cruz Predatory Bird Research Group, offers citizen scientists an opportunity to participate in one of the world's most successful programs to rehabilitate a formerly endangered raptor.

thereby complementing natural food sources. Of course, not all of us own property or have access to open spaces. If you live in an apartment, a rental unit, or in the city, then a seed feeder, birdbath, or hummingbird feeder is likely to attract birds to your home. This is a good thing, at least for you, but there are a few precautions to consider:

- Keep feeders or baths clean and change the water (or nectar) regularly.
- Place feeders or baths near cover so that the birds have a route of escape in case a predator attacks.
- Make sure those windows that reflect light have stickers or shades that will reduce glare and so reduce, or prevent, window strikes. Windows are a major source of bird mortality.
- Make sure that no house cats have access to the bird-feeding area; house cats are major predators and cause untold numbers of bird deaths every year in every region.
- Be consistent. If you commit to feeding birds, try to keep the feeders well stocked.

It is important to understand that by feeding, you are subsidizing the local avifauna and may be attracting undesirable birds to your property—House Sparrows, Eurasian Starlings, Eurasian Collared Doves, jays—that may compete with, exclude, or even depredate more desirable species. So, feeding birds is an individual decision and should be informed by the circumstances of your home turf.

In the technological age in which we live, recordings of bird vocalizations are readily available and can even be broadcast from your smartphone. These programs are helpful tools for learning to recognize species by ear. However, during the nesting season, in particular, they should be used with extreme caution. Birds sing to advertise their territory and attract mates. When we play the territorial song of a warbler or a thrush within an occupied territory, simply to attract a bird into our field of vision, the bird whose territory we are invading may perceive us as intruders or competitors, and may expend extra energy to fend us off and exert his territorial imperative. Energy expended is energy wasted and may compromise the bird's reproductive success. This precaution is especially

TABLE 2 Counties of the Northern California coast, south to north and total number of bird species recorded as of August 2012.

County	Square Miles	Bird Species	Bird Species/ Square Mile
Monterey	3,771.07	486	0.129
Santa Cruz	607.16	433	0.713
San Mateo	1,304.01	453	0.347
San Francisco	232	458	1.974
Marin	828.20	485	0.585
Sonoma	1,768.23	422	0.239
Mendocino	3,878.14	410	0.106
Humboldt	4,052.22	455	0.112
Del Norte	1,229.75	427	0.347

The smallest county, San Francisco, includes the closely observed Southeast Farallon Island, a magnet for vagrant species.

important regarding the rarer and at-risk species—owls, rails, and many passerines.

Excellent advice on the ethics of using taped vocalizations to attract or view birds is given on David Sibley's website Proper Use of Playback in Birding. It is too long to summarize here, but the primary guidance is to exercise respect and be courteous to the bird (and other birders) by limiting the volume and frequency of the playback.

The American Birding Association (ABA) has developed a document, *ABA Principles of Birding Ethics*, that addresses the issue of ethical, or respectful, nature watching quite thoroughly.

The coast of Northern California hosts a remarkable diversity of avifauna and arguably offers the most exciting opportunities for the field ornithologist in North America. The total number of species recorded to 2012 for each county (table 2) gives the reader an idea of just how diverse a region it is.

FAMILY AND SPECIES ACCOUNTS

Ducks, Geese, and Swans

Family Anatidae

Many members of this diverse group of waterfowl are familiar, even to the most casual observer. Most people can distinguish swans, geese, and ducks generically. As you become more familiar with waterfowl, the remarkable variety within the group becomes ever more evident. But first, let's discuss the characteristics the waterfowl share. All members have heavy bodies, short tails, and broad wings. Wide, webbed feet adapted for swimming can be folded on the forestroke and opened on the backstoke to propel the birds through the water. Strong, laterally compressed legs can slice through the water with little resistance. The inner edge of the bill is lined with a small ridge of serrations (lamellae) that can function like a sieve to extract food from the water or mud, or strip seeds from vegetation.

The differences among the subfamilies and tribes within the family Anatidae can be generalized, but each species has its own unique complement of characteristics. The geese and swans are lumped into one subfamily (Anserinae), but each group is placed in its own tribe.

The geese (tribe Anserini) are primarily grazers and have blunt, relatively small bills for foraging on both terrestrial and aquatic vegetation. They tend to be large bodied with longish legs placed at the center of gravity, which enable them to easily walk on land. Geese are monomorphic, with male and female plumages identical, but sexes within a species can vary in size.

The swans (tribe Cygnini) are grazers, but they also tend to forage on the margins of ponds, and their elongated necks allow them to reach the bottom to siphon out food items—submerged vegetation, seeds, and benthic invertebrates. The bill is broad and deep at the base, with nails on the edges of the upper and lower tips—a sturdy tool for prying tubers and roots from the mud or for gathering seeds.

All the "true ducks" are included in one subfamily (Anatinae) and four distinctive tribes—the dabbling ducks (tribe Anatini), the bay ducks (tribe Aythyini), the sea ducks (tribe Mergini), and the stifftails (tribe Oxyurini). All but the dabbling ducks are commonly referred to as "diving ducks." Of 35 species in North America, 24 species occur each winter on the Northern California coast. Most of these exhibit distinct sexual dimorphism, with the males and females distinguishable—males wear spectacular patterns and colors most of the year, and females are less colorful, dressed in subdued earth tones. (While incubating eggs, it is important to be camouflaged from potential predators.) Most ducks have broad, flattened bills adapted to a particular diet, but there is a wide variety in bill shape, ranging from the broad spatulate bill of the Northern Shoveler to the thin, scissorlike bill of the mergansers.

Greater White-fronted Goose *Anser albifrons*

LENGTH 29–35 IN.

White-fronted Geese, as well as Snow and Ross's Geese, are much more abundant at inland valleys in winter, but every year a few of each occur along the coast. Small groups of White-fronts—called "speckle bellies" or "specks" by hunters—probably family groups, may show up anywhere in September or even August but are more common during the waterfowl influx period, mostly in November. This is an Arctic nesting species with a circumpolar distribution. (The Pacific population is composed of two populations with unclear subspecific nomenclature.) Sometimes wild coastal White-fronts "move in" with domestic geese or ducks and never leave.

SPECIAL STATUS Tule Greater White-fronted Goose is considered a Bird Species of Special Concern (wintering), priority 3.

Emperor Goose *Chen canagica*

LENGTH 25 IN.

The beautiful and trustful Emperor Goose is a rare winter visitor to coastal Northern California, but when one or a very small group of them appear, they usually find harbors, estuaries, boat launches, or farm ponds to their liking but remain for only short periods, never overwintering. Emperor Geese will loosely associate with Brant or Cackling Geese. Unlike their congeners the Snow and Ross's Geese, Emperors do not associate with domestic barnyard geese.

Snow Goose *Chen caerulescens*

LENGTH 28–30 IN.

Ross's Goose *Chen rossii*

LENGTH 23 IN.

The Snow Goose and Ross's Goose are much more abundant at inland valleys in winter, but every year, usually during spirited weather in November, mixed flocks "overshoot" their final destinations and swirl along the coast. Noisy skeins, often containing dozens of white geese, may be seen flying between coastal estuaries and reservoirs, or grazing in open pastures amid Canada Geese. The Snow Goose is the more common of the two, but a few Ross's appear at favored locations every year. Both species are reported on most coastal Christmas Bird Counts, with higher numbers to the north. Like White-fronts, a few Snows or Ross's may settle in with domestic geese or ducks and stay for long periods, even years. Both species have a very rare dark plumage variation ("blue-morphs") that superficially resembles the Emperor Goose.

Brant *Branta bernicla*

LENGTH 25 IN.

Pl. 13

The Brant (or Black Brant) is the smallish sea goose of those coastal bays and estuaries that are shallow enough to support beds of eelgrass *(Zostera marina)*—the Brant's favorite food. Few native waterfowl rely so heavily on a single type of plant for sustenance; however, Brant do also graze on marine algae, salt-marsh plants, and upland grasses (Reed et al. 1998). Though they were hunted to very low numbers by the 1950s, better protection has allowed a population rebound. Especially in spring, strings of Brant can be seen migrating offshore, low against the horizon, usually in groups of several dozen but sometimes many more. Hundreds or even thousands gather amid eelgrass beds, especially in April, to refuel en route from their wintering grounds on the west coast of Baja to their nesting territory in the Arctic. These foraging flocks are often quite chatty, braying a soft and guttural conversation among themselves.

Brant are also rather skittish, even for geese, perhaps because of a long history of hunting pressure. They will take flight en

masse in response to the approach of a boat, an overflying aircraft or eagle, or other sources of disturbance. The fall migration corridor, taking the birds on their 3,000-mi journey from Alaska to Mexico, is much farther offshore than the spring route. After staging and fattening up in Izembek Lagoon, Alaska, most fly *nonstop* over the open ocean, but small flocks stop briefly at Humboldt Bay and other large, protected embayments. Small numbers overwinter. The best places to see Brant on the Northern California coast are Drakes Estero, Tomales Bay (northwest of Hog Island), Bodega Harbor, and Humboldt Bay (south).
SPECIAL STATUS Bird Species of Special Concern (wintering, staging), priority 2.

Canada Goose ***Branta canadensis***
LENGTH 28–50 IN.
Pl. 7

Cackling Goose ***Branta hutchinsii***
LENGTH 24–29 IN.
Pl. 14

Canada Geese have always been winter visitors to the coast; it is only since the 1980s that some of them have become permanent residents, and many now raise their young here. There are four easily recognized forms in winter, and each has its own story. In 2004 the Canada Goose complex was split into two species, Canada Goose (the big ones) and Cackling Goose (the small ones), leaving medium-sized ones in between.

The Great Basin Canada Goose *(B. c. moffitti)* is the largest (45–50 in.) and most common, often abundant and rather tame. Natural nesting is in the Sierra Nevada and the Cascade Range, but transplants by humans have become feral and are year-round coastal residents. These largest western Canada Geese are also called "honkers" by hunters.

The Lesser Canada Goose *(B. c. parvipes)* is very similar to the honker but is noticeably smaller (28–30 in.) in all regards. This is the rarest of the four Canadas to reach the outer coast.

The Aleutian Cackling Goose *(B. hutchinsii leucopareia)* is the larger (27–29 in.) of the Cackling Geese. It was one of the very first animals to be placed on the federal list of endangered species. Introduced foxes at its Aleutian Islands nesting grounds had reduced the goose population to about 700 individuals. Fox removal and hunting limitations led to the recovery of this fine bird, and by the year 2000, there were 37,000 counted in Del Norte and Humboldt Counties alone. By 2005, the overwintering population was estimated as 64,000 birds, and spectacular concentrations now enliven north coast bottomlands in March and April. Individuals and sometimes flocks are occasional visitors farther down the coast.

The "small" (24–27 in.) Cackling Goose *(B. hutchinsii minima)* is a small replica of the honker and is about the same size as a Mallard. While similar in pattern to other white-faced geese in this complex, this goose is darker bellied and has a very small bill. It is still common at places in the Central Valley, though the population has been in decline. A few Cacklers are sprinkled along the coast each winter, often among flocks of larger and more numerous Canada Geese. Cacklers may also mingle with domestic geese and ducks.

Tundra Swan ***Cygnus columbianus***

LENGTH 50–55 IN.

Formerly called Whistling Swan, the Tundra Swan is generally scarce along the coast, but small flocks, often in family groups, may put down at any calm body of water, even farm ponds, or damp bottomlands. The most reliable places on the coast to observe native swans—sometimes in the hundreds—are near the mouth of the Garcia River just north of Point Arena in Mendocino County and the Humboldt Bay National Wildlife Refuge between November and February. (See the Point Arena information in Birding Opportunities and Roadside Nature Centers.) The swans graze with various ducks and geese on pasture grass or on aquatic vegetation in water up to about a 3 ft in depth.

Similar Feral Mute Swan *(Cygnus olor)* individuals *may* show up anywhere, but they are most likely in Marin and Sonoma Counties, where a free-flying population has become established. Although native to Europe, escapees have formed wild, permanent groups here and elsewhere in North America. The Mute Swan has a longer tail than does the Tundra Swan and holds its neck in an S-shaped curve rather than relatively straight. When this swan is swimming, its wings are folded higher than on other swans, and the adult Mute Swan has a red, not black, bill. The immature Mute Swan has a completely dark bill and is easily mistaken for a Tundra Swan on first blush. A swan that is vocalizing is not a Mute Swan.

Wood Duck ***Aix sponsa***

LENGTH 18 IN.

Pl. 15

This is one of the seven species of North American ducks that nest in tree cavities—old woodpecker holes, natural cavities, or nest boxes—usually located directly above water. Ponds, streams, or slow-moving rivers with overhanging riparian vegetation along the shore are Wood Duck habitats. The striking pattern of colors of the male and cryptic beauty of the female cause this to be a favorite waterfowl of people looking for natural beauties. Both sexes have crested heads. In flight—usually in small groups—the Wood Duck's wing beats may issue a soft whirring sound. The rather long, rectangular tail and the soft, whistled flight calls are distinctive. In fall, numbers may gather to forage on acorns in oak woodlands that border quiet watercourses.

Wood Ducks are favored by hunters and were nearly exterminated by predatory humans in California by the 1930s, but the population has responded favorably to recent protection and nurturing. Enforcement of hunting regulations, habitat enhancement, and numerous organization-based nest box programs have returned populations to relatively healthy levels.

Gadwall ***Anas strepera***

LENGTH 21 IN.

Pl. 16

The Gadwall is a common year-round resident of fresh and brackish wetlands, with numbers increasing in the winter months as northern and inland breeders arrive to swell the local population. Fewer Gadwalls than Mallards nest here, but they may outnumber Mallards at favored places in winter. Compared with most other ducks, the Gadwall is plain, but close views of the adult male reveal a fine reticulated pattern over much of the plumage. The female may be easily confused with the female Mallard, but she is smaller, and the angle to her forehead is steeper than the gentle slope of the Mallard. She also has conspicuous orange edging to the upper mandible, and she flashes a white speculum pattern in flight.

The Gadwall has been increasing throughout its North American range in recent decades, a trend attributed to a number of factors: a relatively late nesting season that decreases competition; a preference for nesting on islands, decreasing mammalian predation; and changing rainfall patterns on its primary nesting grounds in the Great Plains and Canadian Prairie Provinces.

Eurasian Wigeon ***Anas penelope***

LENGTH 20–22 IN.

Arrival and departure for its winter residence are identical to that of the American Wigeon, but the Eurasian Wigeon is much rarer. Expect to find one or more males (most females are nearly indistinguishable from American Wigeon females) wherever American Wigeon congregate. The Eurasian drake, with his bright crimson head and gray flanks, stands out amid the flock. At favored locations (such as Bolinas Lagoon and the Arcata Bottoms) it is possible to see multiple individuals on a single day. Eurasian Wigeon wintering here arrive in North America from Siberia. Once here, most do not return to their native continent but pair off with their American Wigeon cousins; hybrids between the two species are occasionally detected.

American Wigeon ***Anas americana***

LENGTH 20–22 IN.

Pl. 17

The American Wigeon, also known as "baldpate" for the distinctive white forehead and crown of the male, is a common winter visitor to fresh and brackish wetlands along the coast. At many coastal sites it is the most common wintering waterfowl. Most arrive in August and September and stay here until late April. This is the most northerly nesting dabbling duck, with a breeding distribution that ranges from the shore of the Bering Sea south to the northern prairies. It is a surface-feeding duck, primarily vegetarian in winter, eating aquatic plants, but sometimes foraging in flocks on upland grasses and forbs.

The plumage varies widely, but the male is distinctive with his "bald pate" and a brilliant iridescent green swath surrounding the eye and sweeping back toward the nape. The female and the male in eclipse (nonbreeding-season) plumage are duller, but both have a warm cinnamon or pinkish-tan wash on the flanks that is obvious from a distance. Other useful marks are the rounded head shape and the smallish blue-gray bill with a black tip and nail. Like the Pintail, the American Wigeon rises almost vertically when taking flight from the water. Wigeon have a voice unique among ducks, a sound emblematic of our coastal estuaries in winter—a gentle, three-part whistle, the middle element emphasized—that is easily imitated.

Mallard ***Anas platyrhynchos***

LENGTH 23 IN.

Pl. 18

The quintessential "dabbling duck," although larger than most, the Mallard is the most cosmopolitan and perhaps the most recognizable of the waterfowl. The Mallard is a common resident of fresh and brackish wetlands all along the coast year-round, but it is more common in winter than summer. Its overall abundance is a testament to its broad habitat associations and adaptability to human-altered environments. Nests are usually hidden on the ground among dense upland vegetation, typically rather close to the water's edge. The male in full plumage is unmistakable, with the glossy metallic green head (with an iridescent violet cast in bright light), white necklace, and bright yellow bill. The female and juvenile are feathered in cryptic browns and blacks that camouflage them in mud or marshy vegetation. It is the adult female that delivers the loud, descending "quack, quack, quack, quack, quack." Males mutter much more quietly.

Blue-winged Teal ***Anas discors***

LENGTH 15 IN.

Pl. 19

While the body and facial patterns of male Cinnamon and Blue-winged Teal are quite different, their wing patterns are nearly identical, as are the females of the two species. They are very closely related, and hybrids, though rare, are found annually in California.

Although never as common as Cinnamon Teal on the coast, Blue-winged Teal have apparently increased in number, according to winter records of recent decades. The increase may be real or may be the result of more and more competent observers in the field. This species is most often found in quiet ponds and inlets, often paired and in the company of Cinnamon Teal.

NOTE Cinnamon Teal males, females, and young have sky-blue dorsal wing coverts, so the big patch of blue on the upper wing near the body is not a diagnostic field mark for Blue-winged Teal. Northern Shoveler also belongs to this blue-winged group of dabblers.

Cinnamon Teal *Anas cyanoptera*

LENGTH 16 IN.

Pl. 20

Unlike most coastal waterfowl, Cinnamon Teal are more common here in spring and summer than in winter. Most are in Mexico from October through March. The Cinnamon Teal visits fresh or slightly brackish waters with muddy bottoms. The bill is relatively large and noticeably shovel-like, an adaptation for siphoning vegetation and seeds from the pond surface or the mud. The male has a piercing red eye. Cinnamon Teal breed sparingly at favored places along the coast, nesting in dense tules, sedges, or rushes. Much of their forage is derived from ingested (or filtered) organic mud. Cinnamon Teal are seasonally monogamous, pairing up before arriving on the breeding grounds, and so are often seen in pairs here, rarely in flocks of more than a half dozen individuals.

Northern Shoveler *Anas clypeata*

LENGTH 20 IN.

Pl. 21

These rather bulky, large-headed, somewhat top-heavy dabblers are closely related to the teal. The wing pattern of the male is very similar to those of Cinnamon and Blue-winged Teal. Shovelers are widespread and local in winter, scarce and local in summer. The drake is a spiffy bird in alternate plumage, but the female is a rather dull mottled brown, similar to other female dabblers except for her oversized bill.

Shovelers congregate at shallow, mud-bottomed estuary edges, salt ponds, and sewer ponds where the substrate has a high organic component or where aquatic invertebrates swarm. The hunter's nickname, "spoonbill," well describes the peculiar, specialized bill—wider at the tip than in the middle, with comblike serrations (lamellae) lining the lateral edge. Shovelers use this unique apparatus for filter feeding, either while swimming and straining out small invertebrates or by shoveling up and sifting through organic mud. We might think of the lamellae as avian baleen. Shovelers are gregarious ducks, often milling together in large flocks that may form "carousels," spinning slowly on the surface of open water, "vacuuming" small swimming animals (often brine shrimp) toward the surface.

Northern Pintail ***Anas acuta***

LENGTH 22–24 IN.

Pl. 22

The Pintail is a long-necked dabbling duck, elegant and slender in profile, with a long pointed tail. The adult male in definitive breeding (alternate) plumage is a handsome fellow, not likely to be confused with any other species. White stripes extending from the breast up the sides of the neck, converging behind the head, and offset by the chocolate brown of the face and hindneck create a most distinctive, and recognizable, pattern. The Pintail is graceful and agile in flight, longer and thinner, more pointy front and back, than any other duck. On takeoff, it is able to spring upward almost vertically from the surface of the water, a trait that has earned it the hunter's nickname "sprig." Pintails are strong fliers, often venturing out over the open ocean.

While most dabblers are strictly surface feeders, when the Pintail is on the water, it often tips forward, fully submerging the head, neck, and breast so that the rear end is sticking straight up. An exceptionally long neck allows it to forage on the bottom and glean food items not readily available to other dabblers. This is a gregarious species in winter, gathering mostly with its own kind.

Pintail are common winter visitors to the coast, with most

arriving in late August and September, departing in March and early April. In some coastal lagoons and estuaries, the timing of departure depends on cumulative rainfall; as seasonal wetlands become hydrated, Pintail may abandon the coast for newly available habitat inland. August arrivals are all in female-like cryptic plumage, with some individuals ocher in color. A few Pintail summer along the coast, and there are occasional nesting records from coastal counties, but the core nesting habitat is in the interior.

Green-winged Teal — *Anas crecca*

LENGTH 13 IN.

Pl. 23

This is our most compact dabbling duck, so diminutive that according to Leon Dawson (1923), it "takes two birds on a single plate to provide a meal for a hungry man." Indeed, this is one of the most common ducks taken by hunters, which may explain its nervousness when approached by humans. Nevertheless, numbers may be increasing nationwide.

Green-wings favor grassy freshwater pools, flooded bottomlands, and shallow back edges of estuaries where salinity is somewhat lower. Green-wings eat aquatic vegetation near the surface and derive some protein from foraging along shores. Invertebrates such as worms, snails, and small crustaceans are taken opportunistically. These are usually the most common ducks seen foraging on mudflats with shorebirds.

In addition to the small size and rounded head profile, the vertical white stripe on the foreflank of the male is an obvious field mark, even at a distance. In flight, these small dabblers are quite nimble, with many feints and turns. Green-wings usually form tight flocks that twist and turn in unison, almost as choreographed as shorebird flocks. The wings produce a soft whistling sound.

Individuals of the recognizable Asian subspecies *(A. c. crecca)* are occasionally found along the California coast in winter—almost always with their American congeners. The male of this subspecies has a white horizontal stripe down each side of the back (scapulars) and lacks the vertical, white half crescent at the front of the flank.

Canvasback *Aythya valisineria*

LENGTH 21–23 IN.

Pl. 24

Considered "king of the pochards" (bay divers in the genus *Aythya*) by hunters and birders alike, this elegant species is abundant locally along the coast in winter. Huge flocks forage or raft together on the larger bays, and small numbers gather at lakes or ponds. San Francisco and San Pablo Bays support major wintering concentrations for this duck, which nests primarily on the Canadian prairie. Canvasbacks occur only in North America, and they are monotypic, that is, there are no subspecies.

This is a sturdy, large-bodied diving duck that forages on the surface or dives as deep as 30 ft. In profile, the head shape is distinctly wedge shaped because the long bill, very deep at the base, slopes gently up to the forehead. This head shape helps distinguish the Canvasback from the somewhat similar Redhead in any plumage. (The Redhead has a more rounded head and a shorter, less broad-based bill.) Canvasbacks nest at marshy lakes and potholes throughout northwestern North America, north of California. Look for them here from October into March.

Redhead *Aythya americana*

LENGTH 19–21 IN.

Pl. 25

This handsome diving duck once nested in the coastal counties of Alameda and Monterey (Grinnell and Miller 1944), but habitat alteration and hunting reduced the numbers through the 20th century. Now, nesting is confined to the Central Valley and the northeastern portions of the state, and the Redhead is an infrequent winter visitor to the coast. Although similar to the Canvasback, this species has a rounder head and a tricolored bill, and the male has yellow, not red, eyes. A careful search of a large raft of divers—especially Canvasbacks and scaup—may reveal a few Redheads mixed in. Small numbers also winter in constructed lagoons, sewage ponds, and reservoirs.

In the west, this species nests mostly at prairie potholes and is dependent on wet years for breeding habitat. Populations swing drastically with wet and dry cycles. Agencies regulate hunters' bag limits (the number of birds that may be killed) when Redhead numbers are low.

SPECIAL STATUS Bird Species of Special Concern (breeding), priority 3.

Ring-necked Duck *Aythya collaris*

LENGTH 18 IN.

Pl. 26

This dapper diver is superficially similar to the scaup, but the male has a darker (black) back and an obvious vertical white line, immediately in front of the bend of the folded wing, separating the dove-gray flanks from the black breast. All Ring-necks have tall, squared crowns and decorative bills—grayish with a narrow white band distally, and a black tip. The dark brown "ring neck" is rarely visible in the field; a more descriptive name might have been "Ring-billed Duck." The female has a distinctive white eye-ring and a white line extending back from the eye. Only the smaller female Wood Duck might be confused with the hen Ring-neck.

Shallow freshwater lakes and ponds, often right on the coast, are the favorite wetlands of Ring-necks. They rarely visit salt water.

Greater Scaup *Aythya marila*

LENGTH 17–19 IN.

Lesser Scaup *Aythya affinis*

LENGTH 15–17 IN.

Pl. 26, 129

These two scaup, called "bluebills" by hunters, have very similar plumages but can usually be distinguished by head shape, wing pattern, and sometimes habitat preference. The head of the Greater is large and, in profile, evenly rounded; that of the Lesser is smaller and slightly squared, or angled at the back of the crown. In ideal light, the iridescence of the dark head of the male may be a helpful clue to identity: greenish in the Greater, a purplish cast in the Lesser. Also, the black "nail" on the end of the bill is more obvious on the Greater than on the Lesser Scaup. The white flash in the flight feathers of the

Greater extends the length of the wing, whereas the Lesser has white only on the inner wing (secondaries).

Greaters occur almost exclusively in marine waters, and Lessers tend to frequent fresher waters. In rafts of scaup on coastal estuaries and bays, Greaters far outnumber Lessers. Scaup found away from the coast are mostly Lessers. When both species are present, Lessers usually forage in fresher portions of an estuary or in shallower waters close to shore—giving a misleading impression of the true ratio.

The Greater Scaup is the only species of the bay duck group, that is, genus *Aythya*, with a circumpolar distribution. Greaters congregate in impressive numbers in the larger coastal embayments, especially during winter herring runs. San Francisco Bay is a major wintering ground on the Pacific coast. Both species are rare in summer, but Lesser Scaup nests very occasionally around San Francisco Bay.

SEA DUCKS

The "sea ducks" of the tribe Mergini include the three species of mergansers and the goldeneye genus *(Bucephala)* plus the less closely related scoters *(Melanitta)*, Long-tailed Duck *(Clangula hyemalis)*, and Harlequin Duck *(Histrionicus histrionicus)*.

Harlequin Duck *Histrionicus histrionicus*

LENGTH 15–16 IN.

Beauty combined with rarity causes a Harlequin Duck sighting to be highly coveted by birders and anyone else who has an interest in nature's beauty. The stunning pattern and hue of the adult male is unsurpassed within the waterfowl. The female Harlequin resembles the female Surf Scoter but is more petite, with a rounder head.

Harlequins are hypercoastal in winter, with small numbers found in intertidal and subtidal waters along rocky coastlines or amid kelp forests on the outer coast and larger estuaries. The farther north you look along our coast, the more likely you are to find a Harlequin. Point St. George in Del Norte County supports the highest numbers, with abundance diminishing southward. Harlequins forage for marine invertebrates, usually in relatively shallow water.

Harlequins formerly nested along turbulent rivers and streams in the Sierra Nevada and the Cascade Range, but Northern California is at the southern edge of the breeding range, and numbers have dropped over the last century. Now, the few we see are probably from source populations in Canada or Alaska.
SPECIAL STATUS Bird Species of Special Concern (breeding), priority 2.

Surf Scoter *Melanitta perspicillata*

LENGTH 20–23 IN.
Pl. 27

The most abundant sea ducks along the coast, Surf Scoters raft in winter off most ocean beaches, just behind the rolling breakers, and flock inside major embayments. The nearshore migration—north in spring, south in fall—seems endless, as line after line of scoters (sometimes in V-shaped formations) pass daily for months, flying low, just above the ocean swells.

In the distance, Surf Scoters just look like big black (male) and brown (female) ducks. Close up, the bill of the male is a brilliant red, orange, and white, very deep at the base, strengthened by a dime-sized, blue-gray plate on each side that is used for snipping off mollusks underwater. The large white patch on the nape of the adult male is unmistakable. In spring the young males molt from drab winter plumage and appear in a wide variety of dress as they progress into adult breeding plumage. The wings of the Surf Scoter are fairly large relative to body size, allowing it greater agility in flight than most diving ducks have. Although the Surf Scoter is quite silent, the wings whistle loudly in flight, especially on takeoff.

White-winged Scoter *Melanitta fusca*

LENGTH 22–25 IN.

A stocky sea duck, the White-wing is most likely to be seen near shore in the breakers or in a raft with other divers—scaup and scoters. Both sexes have a white wing patch (speculum), usually visible in the field, and a swollen base to the bill that shows a distinctive profile. The male is velvet black with a small white apostrophe surrounding the eye. The female and young male are similar to the Surf Scoter, dusky brown with two whitish patches on the face, placed low in front of and behind the eye. This is the largest of the three scoter species; its size and the fact that it rides lower in the water than the Surf Scoter may help with identification.

Much less common than the Surf Scoter, the White-wing's presence in winter along the California coast seems to have decreased since the 1950s. This may reflect a broad population decline or a cyclical downturn—possibly due to some local variable. White-winged Scoters forage over sandy, nearshore shallows of bay and ocean waters, feeding on bottom-dwelling marine mollusks and crustaceans. The Mole Crab *(Emerita analoga)* is a favored food. Mole Crabs are intermediate hosts to a parasite group called "thorny-headed worms" (Acanthocephala); sea ducks (eiders and scoters) are the definitive hosts. Once ingested, the worms develop in large numbers and get energy to grow and reproduce by devouring the digestive system of the duck.

Black Scoter ***Melanitta americana***

LENGTH 16–18 IN.

Pl. 28

This is the smallest of the three scoters, with a rounder head and more delicate bill. Distant males may remind you of a coot, distant females of a winter Ruddy Duck. Like the White-wing, the Black Scoter is relatively scarce in the southern portion of our region, becoming more common northward. Drakes Bay at Point Reyes is the southernmost locality of regular occurrence. Foraging is usually near shore, where rocky intertidal or wharf piling habitats occur. Black Scoters also raft with Surf Scoters on open water for preening and resting.

Long-tailed Duck ***Clangula hyemalis***

LENGTH 15–16 IN.

Formerly known as Oldsquaw here, this bird has always been called Long-tailed Duck in English-speaking Europe, in reference to the two elongated central tail feathers of the male in breeding plumage. This bird breeds in the Arctic, though, so most individuals that reach Northern California have short tails in fall and winter. Even so, the male is unmistakable with a largely white head and a dark cheek patch. The winter female is also mostly white, but the plumage of this duck varies more than any other.

Rather rare, and always an exciting find at this latitude, Long-tailed Ducks are usually seen singly or mixed in with scaup or scoters at bays and harbors. Males have a pink "saddle" across the diminutive bill; females, a gray one. This is a vocal sea duck: the genus name, *Clangula*, refers to the clanging vocalizations, a loud "ow ow owoolik."

Bufflehead ***Bucephala albeola***

LENGTH 12–14 IN.

Pl. 29

The smallest diving duck, this beautiful little charmer is a cousin of the goldeneyes. Buffleheads nest in northern forests, usually in cavities excavated by Northern Flickers. On the breeding grounds, they forage in freshwater ponds and lakes. In winter they switch to coastal marine waters and become common to abundant in bays and harbors as well as lakes and ponds all along the coast. Large, protected estuaries (Crescent City Harbor; Humboldt, Bodega, Tomales, and San Francisco Bays; and Bolinas Lagoon) are particularly favored by large numbers of this species. During a herring run, and for much of winter, as many as 10,000 Buffleheads will concentrate on Tomales Bay alone. Herring runs occur in the last part of December and throughout January with varying degrees of intensity year to year. Like the also abundant scaup and scoters, Buffleheads do not prey on the herring, but rather on the roe that is attached to submerged vegetation. They also eat small crustaceans and mollusks.

Coupled with its diminutive size, the black-and-white plumage of the male is unmistakable. The female is similar in size and shape to the male, but her plumage is mostly gray brown with a small, distinctive white patch behind and below the eye.

Common Goldeneye *Bucephala clangula*

LENGTH 18–19 IN.

Pl. 30

Commons are locally common in midwinter at favored estuaries and sheltered brackish hideouts like constructed lagoons and impoundments, south of Point Arena. After nesting in the northern boreal forests, goldeneyes do not arrive in our region until late in October, the last of the regular wintering ducks to appear. They are found along the coast for the length of the state, but unusually high aggregations occur most years in San Francisco Bay at the mouth of the Petaluma River, where up to 1,000 goldeneyes forage with scaup and other divers. In flight, the wings of goldeneyes make a high-pitched whistling sound, hence the nickname "whistler" from hunters and others.

Both Common and Barrow's Goldeneye males have complex courtship displays, performed here on the wintering grounds, involving elaborate head turning, thrusting, and pumping movements that flash the white facial patterns at the females—truly a performance to behold. The species name of the Common, *clangula*, is the same as the genus name of the closely related Long-tailed Duck *(Clangula hyemalis)*, a much rarer species in our region. *Clangula* is the Latin diminutive of *clangor*, meaning "noisy," and may refer to the vocalizations of the males of both species during courtship display.

Barrow's Goldeneye *Bucephala islandica*

LENGTH 18–19 IN.

Pl. 30

This lovely cousin of the Common Goldeneye is quite rare along the coast (inland, too) except at favored places in the San Francisco Bay Area. At the "yachting" lagoons of Foster City in San Mateo County and Lake Merritt in Oakland in Alameda County, Barrow's Goldeneyes sometimes even outnumber Commons. The scientific species name of the Barrow's, *islandica*, refers to Iceland, where this bird was first "discovered" by curious Europeans as they journeyed forth to explore the world.

Identification of the adult male is aided by the vertical black stripe on the side extending from the shoulder toward the waterline. The female can usually be distinguished from the

Common Goldeneye by her rounder head and butter-yellow bill; head shape is useful clue to identification, less peaked in Barrows than in the Common.

SPECIAL STATUS Barrow's Goldeneye has been extirpated as a breeding species from California and therefore is included on the California Department of Fish and Game list of "Special Animals" (2011).

Hooded Merganser ***Lophodytes cucullatus***

LENGTH 16–19 IN.

Common Merganser ***Mergus merganser***

LENGTH 21–28 IN.

Pl. 31

Red-breasted Merganser ***Mergus serrator***

LENGTH 22–24 IN.

Pl. 32

Of the three California mergansers, the Red-breasted, visiting in winter, is the only one that favors saltwater habitats, and it is rarely seen away from the marine environment. Both sexes sport bushy crests, reddish in the female, black in the male. Not as gregarious as other ducks, the Red-breasted Merganser is most often found in groups of one to five. In a truly unique behavior, they sometimes team with Snowy Egrets *(Egretta thula)* to herd and eat small, schooling fish. The mergansers push the fish toward shore within jabbing reach of the egrets, and the egrets in turn keep the fish offshore, deep enough for the ducks to capture underwater.

The Common Merganser is very similar to the Red-breasted in appearance but associated with freshwater—reservoirs, lakes, rivers, and streams—sometimes venturing downstream into brackish estuarine water. The female Common Merganser is crested, but she differs from the female Red-breasted in having a conspicuous white throat. Common Mergansers nest throughout the coastal counties, and numbers have apparently increased in recent decades for reasons unknown, but perhaps because of an increase in reservoirs.

The Hooded Merganser, the smallest and most elegant of the tribe, is a retiring bird of quiet freshwater ponds and backwaters. It nests in tree cavities or stumps only in the northernmost counties of our region. Small groups of mixed sexes gather on still ponds to display their elaborate courtship behavior in spring—an enchanting ritual to observe, for those lucky enough and stealthy enough. Hooded Mergansers tend to flush readily when approached.

Ruddy Duck ***Oxyura jamaicensis***

LENGTH 15 IN.

Pl. 33, 34

During most of the year, Ruddies are cryptically dressed in browns and tans, but in early spring the males change to a flashier plumage. In this breeding plumage, body feathers are a gingery cinnamon that contrasts sharply with immaculate white cheeks. Erectile feather tracts on the crown look like small horns at certain angles. The bill becomes bright, sky blue. The Ruddy is unique among the North American ducks (subfamily Anatinae) in acquiring its bright (alternate) plumage in spring and summer; other ducks molt into theirs in fall and winter, attaining a more cryptic plumage that provides camouflage during the nesting season.

The male's courtship antics are entertaining. In the "bubble display," the drake cocks his tail and holds his head erect, then begins slapping his breast with his bill, forcing air out of his plumage to form bubbles in the water. This performance ends with an odd vocalization, a rapid succession of rattles followed by a low belch. Ruddy Ducks nest in the vegetated edges of ponds and in fresh to brackish marshes throughout the region, but in sparse numbers. They are abundant on some estuaries during winter, rafting together or intermixed with other divers, especially Buffleheads. San Francisco Bay is an important wintering ground.

Loons

Family Gaviidae

Each of the five species of North American loons sport spectacular breeding (alternate) plumage, but most individuals that winter along the Northern California coast are in drabber winter (basic) dress. All are large, heavy-bodied birds that tend to ride low in the water. Loons fly with strong and steady wing beats—necks outstretched, feet extending out behind. In direct flight, the head is held low, below the body, giving a hump-backed profile. All are diurnal migrants. Loons are fish eaters, superb underwater swimmers propelled primarily by their large webbed feet on strong legs placed far back on the body. When diving, they can stay submerged for minutes at a time and may travel some distance beneath the water, surfacing at distances unexpected and difficult for the observer to predict.

Three species—Red-throated Loon *(Gavia stellata)*, Pacific Loon *(G. pacifica)*, and Common Loon *(G. immer)*—occur regularly along the Northern California coast. Two others—Arctic Loon *(G. arctica)* and Yellow-billed Loon *(G. adamsii)*—are quite rare here. The Common and Yellow-billed Loons constitute a superspecies pair, as do the Arctic and Pacific Loons.

Red-throated Loon — *Gavia stellata*

LENGTH 21–25 IN.

Red-throats are common from October through April in harbors, estuaries, bays, and the open ocean just beyond the surf. This smallest of the loons tends to favor shallower water (3 to 30 ft) than the other loons and dives for shorter periods of time. Average time underwater is about one minute (Carboneras 1992). Red-throats are usually solitary, but sometimes hundreds attend foraging frenzies in sheltered bays during herring spawns. Very few remain in the region through summer. The forecrown is sloped, and the bill, thin for a loon, is noticeably angled upward and is usually carried above horizontal. Crown and nape are pale gray, white surrounds the eyes, and the flanks often show white along the waterline. The Red-throat is most similar to the Pacific Loon, which has a thicker neck, straighter bill, and darker nape and sides. Red-throats tend to be silent while wintering in California.

Pacific Loon — *Gavia pacifica*

LENGTH 25–29 IN.

Pacific Loons are abundant as coastline migrants and common winter residents on large bays and nearshore ocean waters.

Migration is most impressive in spring (April through June) when hundreds of thousands, most in crisp breeding plumage, stream past coastal promontories, line after line, day after day (mostly in morning hours), all northbound. Massive movements can also occur in late fall and early winter (November and December) just offshore, with thousands gathering where convergent currents concentrate prey. Small numbers oversummer, most in winter plumage. (It takes as much energy to molt as it does to migrate, so birds that "miss" a migration may also "miss" replacing their plumage, retaining winter plumage during summer.)

Pacific Loons occur in mixed-species flocks more commonly than do other loons, often foraging actively in the company of Brant's Cormorants. In profile, the neck is rather long and thick, with distinct contrast between the gray hindneck and white throat. The Pacific Loon has a straighter, stiletto-like bill compared with the other species. Typically, the bill is carried parallel to the water's surface. (With an ideal view, the thin "chin strap" characteristic of this species might be visible.) When diving, this species is apt to lunge forward, unlike the Common and Red-throated Loons with their subtler sinking dives. In flight, the Pacific Loon holds the head and neck straighter, more in line with the axis of the body, than do other loons. In breeding plumage and in superb light, the Pacific Loon may show greenish throat sheen, a plumage characteristic also present in the Siberian race of the Arctic Loon *(G. a. viridigularis).*

Arctic Loon ***Gavia arctica***

LENGTH 25–29.5 IN.

Very similar in appearance to the Pacific Loon and extremely rare in Northern California, the Arctic Loon was recognized as a species distinct from the Pacific Loon in 1985 (AOU 1985). The "Green-throated" Arctic Loon *(G. a. viridigularis)* is sympatric with the Pacific Loon on nesting grounds in western Alaska and eastern Siberia. The few occurrences in Northern California have been of single birds, and identification is nuanced and difficult (Reinking and Howell 1993), but the Arctic Loon may be distinguished from other loons by an extensive white patch on the flanks at the waterline.

Common Loon *Gavia immer*

LENGTH 29–35 IN.

Pl. 8, 9, 35

Usually solitary, but sometimes occurring in small groups of two to several individuals, this largest of the three regular species is rather common from October through April in harbors, estuaries, and bays and along ocean beaches just beyond the breaking surf. Small numbers remain over the summer months, dressed in winter (basic) plumage. In spring, before migrating north to the breeding grounds, some adults may be seen in their fancy breeding (alternate) plumage. Then the "necklace" is evident, the subject of many stories and myths of the native Pacific coast people. Males tend to be slightly larger than females, but plumages are similar. There are no subspecies (monotypic). The haunting, quavering yodel of the male may be heard at twilight in October and April from quiet coves along the coast. Common Loons have massive bills with angled lower mandibles, large rectangular heads, and bulbous forecrowns. Bills are carried parallel to the water's surface, except when searching for fish, their primary food. They often paddle slowly along the surface with the head held at a downward angle, looking beneath the surface for prey. To dive, the bird expels air, decreases buoyancy, and sinks smoothly beneath the surface.

Common loons formerly nested in mountain lakes of northeastern California but have not been documented since the 1940s (Grinnell and Miller 1944).

Yellow-billed Loon *Gavia adamsii*

LENGTH 36 IN.

The Yellow-billed nests farther north than the Common Loon, but the winter ranges of the two species overlap in the Pacific Northwest, though rarely as far south as Northern California. The Yellow-billed is very rare on the north coast, not seen some years, though in other years several may arrive to winter locally. Bulkier than the Common Loon, the Yellow-billed is distinguished by an extremely large, light-colored bill, much paler basic plumage, and the habit of carrying its head and bill tilted upward rather than

Common Loon swimming. Some diving birds use the wings to propel with, but this Common Loon, whether swimming on the surface (above) or diving (below), is the consummate foot-propelled diver. The loon can generate remarkable power with lateral strokes of its large webbed feet, placed at the stern of the body. The legs are compressed laterally and the body streamlined to cut through the water and reduce resistance. *K.H.*

parallel to the water. The AOU (1983) considers the Yellow-billed and Common Loons a *superspecies* complex.

Grebes

Family Podicipedidae

Resembling loons superficially, the grebes represent a distinct evolutionary line, although their taxonomic relationship to other waterbirds is not clear, but recent studies found the family more closely allied with the flamingos (Hackett et al. 2008) than the loons! If true, the similarities in loon-grebe morphology provide a striking example (like the penguin-alcid similarity) of convergent evolution and the powerful selective forces on distantly related species.

Unlike loons, all of which have the same basic body model, grebe species vary in size and shape and occupy a wider array of habitats. Grebes are a variable group of perky swimming birds that tend to dive headfirst from the surface. In underwater pursuit of prey, the wings are held close to the body, which is propelled by a full-thrust kick of oar-like legs set far back on the torso and having widely lobbed, not webbed, toes. However, some, such as the diminutive Pied-billed Grebe, may sink, submarine-like, by compressing the lungs. Sexes are very similar in all plumages, although males of some species (Western and Clark's Grebes) may have noticeably larger bills than females. Fish are the most frequent prey of the larger species, but smaller species also take invertebrates and aquatic vegetation.

Pied-billed Grebe ***Podilymbus podiceps***

LENGTH 13 IN.

Pl. 37

The only species of grebe that nests along the coast of Northern California, the Pied-billed is a common year-round resident, as it is throughout much of the coterminous United States. Pied-billed Grebes have a greater preference for freshwater than the other five grebe species, although they also occur on estuarine and even nearshore waters in winter. Their favored habitats are lakes and ponds with marshy edges, and brackish estuaries. One of the characteristic sounds of coastal marshes and ponds is the resonant, wailing "cow-cow-cow . . . " of Pied-billeds emanating from cattails or tules during nesting season. The fortunate observer may see newly hatched chicks, with black-and-white striped heads, riding on the adult's back in a quiet pond or lagoon.

While the other five coastal grebe species are similar in structure and behavior, Pied-billeds are different looking—a basic brown plumage (younger birds have more orangey necks and bills than adults), white "cotton tails," and plump heads.

All grebes can submerge by lunge diving; the Pied-billed can also sink gradually by uplifting its broad tarsi and spreading the toe lobes, expelling air from air sacs and between feathers and body, so that the bird drops below the water's surface, submarine-like. Submerged, it can then chase prey or swim

As with other grebes, the narrow legs of the Horned Grebe are set far back on the body and rotate when the bird is swimming or diving, shown here on the forestroke and backstroke. Also note the toes. Rather than connected webbing (as in ducks), the foot of the grebe has lobes, with flaps along the sides of each toe to form a paddle. *K.H.*

away from a threat and allow only the head to periscope above the water's surface. So, when you see one sink, it may be difficult to relocate the stealthy little grebe.

Horned Grebe ***Podiceps auritus***

LENGTH 13–16 IN.

Horned Grebes do not nest in California but are common along the coast in winter. Rather like a small, dark-billed version of Western Grebe, the Horned have sharply defined white cheeks and foreneck, a flat crown, and a rather short, straight bill, which appears stubbier than that of the thin-billed Eared Grebe. It is frequently found near shore on protected bays, in estuaries, and in small-boat harbors. Numbers sometimes coalesce in foraging flocks and concentrate where herring spawn. They are somewhat more marine in distribution than the similar Eared Grebes. Horned Grebes are absent from the California coast in summer. This species has an elaborate array of pair-bonding ceremonies, but those behaviors are not seen in this region.

Red-necked Grebe *Podiceps grisegena*

LENGTH 17–20 IN.

Pl. 36

Intermediate in structure and size between the large, statuesque Western and Clark's and the small, pudgy Eared and Horned Grebes, the Red-necked Grebe is relatively uncommon and difficult to see along most of the California coast. In profile, the head shape is distinctive, triangular with a symmetrical, straight, and rather thick bill. The gray or whitish cheek patch extending from the bill to the ear is also a helpful mark. Sexes are alike. When they first arrive in fall, many individuals are still in their handsome breeding plumage.

Most often seen in nearshore waters, or near the mouths of the larger estuaries, they are rarely in lagoons, embayments, or sewage ponds. The best places to search from mid-September into April are at the "kelp line" from Monterey Harbor to Point Pinos; off Limantour and Drake's Beaches in Marin County; at the mouths of Bodega Harbor and Tomales Bay; at Point Arena; at the mouth of Humboldt Bay; and at Crescent City Harbor. Red-necks are hypercoastal in winter; they occur in loose flocks, and they become quite rare south of our region. Rarely seen in flight, Red-necks tend to dive when threatened, and they migrate at night.

Eared Grebe *Podiceps nigricollis*

LENGTH 13–15 IN.

Pl. 38

Eared Grebes nest sparingly in California's Great Basin, but it is likely that the many that spend winter along the coast have come from much farther north, perhaps British Columbia or Alberta. The first winter arrivals appear in the last week of September, and the species is common on protected bays, in estuaries, and in small-boat harbors. On occasion, large flocks gather in nearshore waters off headlands. Eared Grebes are the size of Horned Grebes but have smaller heads that are "cresty," "dirtier" cheeks and forenecks, thinner necks, and slightly thinner, upturned bills. These birds often assume a "sunning posture" by fluffing the rear end, giving them an elevated stern, a characteristic helpful to the birder in separating Eared from Horned Grebes in the field. It is interesting to watch such individuals orient themselves with back end toward the sun. Eared Grebes are absent from the California coast in summer but may transition into their spiffy breeding plumage in spring just before leaving the coast.

Clark's Grebe — ***Aechmophorus clarkii***

Western Grebe — ***Aechmophorus occidentalis***

LENGTH 25–27 IN.

Pl. 36, 39

Long-necked and elegant black-and-white grebes, these two common winter species are often seen feeding or rafting in mixed flocks. Although plumages of the sexes are alike, males have noticeably larger and thicker bills than females. Once considered color morphs of a single species, Clark's and Western Grebes were not recognized as separate species until 1985, the result of detailed DNA analysis and behavorial studies (AOU 1985). Although their ranges overlap, both on nesting and wintering grounds, hybridization is apparently rare. Neither species nests on the immediate coast, preferring freshwater lakes to tidal marshlands for nesting. However, in winter, large flocks gather along the coast, in protected nearshore waters, estuaries, and large bays. In such places, many oversummer. Westerns are generally more common (especially from Mendocino Co. north), with Clark's becoming more common farther south. Some authors suggest that Clark's prefer deeper water farther offshore, but we have noted Clark's seeming to prefer shallower sloughs and creeks, as well.

The spring courtship dance, nearly identical in the two species, is an astonishing performance—a combination of Funky Chicken and a graceful synchronized ballet. There is a difference in the advertising call—a double-noted grating "creet-creet" given by the Western, and a similar vocalization but with only a single note given by the Clark's (Storer and Nuechterlein 1992). The dance is sometimes seen in spring in protected coastal embayments such as Lake Earl in Del Norte, Richardson's Bay in Marin, or Elkhorn Slough in Monterey or at outlying nesting reservoirs such as Lake Merced in San Francisco or Lake San Antonio, Monterey County.

Bill color is the best single characteristic for separating the two species—orange for Clark's and greenish-yellow with a dark culmen for Western. Also, the Clark's is whiter overall than the Western, in the face, the neck, and the flanks above the waterline. Both species have bright, fire-engine-red eyes. In the Western, the black of the crown extends down to encompass the eye; in the Clark's it does not, so the eye is within the white facial feathering. Still, even with a good view, variation in plumage is such that some individuals will defy identification. Just wish them well and go on to a more typical example.

Albatross

Family Diomedeidae

Pl. 40
The largest of the tubenoses—indeed, the largest of the seabirds—albatross are legendary ocean wanderers. All members of the family have remarkably long and slender wings, large heads with powerful hooked beaks, short tails, and webbed feet. A half dozen albatross species have been recorded in Northern California waters, but only three species, all in the genus *Phoebastria*, occur regularly offshore in California or in North American waters. Black-footed Albatross *(Phoebastria nigripes)* is the most common of these, and when strong onshore winds guide it shoreward, this magnificent mariner can be seen occasionally from headlands and coastal promontories, soaring above the waves on slender, stiff wings. Laysan Albatross *(P. immutabilis)* is a sister species of the Black-footed, and it rarely occurs inside the continental shelf. Another species, the Short-tailed Albatross *(P. albatrus)*, was fairly common nearshore in the 19th century, but its nesting colonies in Japan were decimated by plume hunters, pushing this oceanic wanderer to the brink of extinction. Fortunately, protection has bolstered the numbers in recent decades and small numbers of Short-tails are now seen annually offshore at Monterey Bay, Cordell Bank, Fort Bragg, and other favored pelagic birding sites (Stallcup 1990, Roberson 2002, Evens and Tait 2005).
SPECIAL STATUS Black-footed Albatross is a Bird of Conserva-

tion Concern in coastal California (USFWS 2008). Short-tailed Albatross is federally Endangered.

Shearwaters and Fulmars

Family Procellariidae

The "procellarids" are true seabirds, and like other members of the order—albatross, petrels, and storm-petrels—they spend most of their lives on the open ocean and do not go to land except to breed. Procellarids are also called "tubenoses" because of their uniquely shaped bills that enable these seabirds to desalinate the seawater they drink and expel impurities through special plumbing, pea-shooter-like tubes on the bill (naricorns or tubenares). There is strong evidence that the tubenoses, especially nocturnally feeding species (e.g., storm-petrels), have a highly refined sense of smell and are able to discern gaseous mists that are produced by phytoplankton, the food of krill. Apparently, these ocean wanderers are following a "smellscape" that leads them to current convergences and upwelling zones, areas of the ocean that produce plankton blooms and, in turn, concentrate zooplankton. Some of the larger diurnal foragers—albatross and fulmars—also use visual cues, such as surfacing marine mammals or anomalies in the sea surface, to find productive feeding areas.

Northern Fulmar ***Fulmarus glacialis***

LENGTH 18 IN.

Pl. 41

Although most tubenoses have affinities with the Southern Hemisphere, fulmars have a more northerly distribution. In flight, fulmars are stockier and broader winged than the shearwaters. These are truly pelagic birds that nest on islands and headland ledges in the Arctic and then spend the rest of the year foraging the North Pacific, usually beyond sight from land, and are numerous only north of about 35°N latitude. Peak numbers in Northern California occur from October through November. There are often autumnal die-offs here near the southern edge

TUBE-NOSED SWIMMERS (PROCELLARIIFORMES)

Four families are included within the order Procellariiformes—albatross, shearwaters, storm-petrels, and diving petrels—the "tubenoses." The name refers to the unique tubular nasal passages atop the bill (naricorns), used for smell and salt extraction, adaptation to life on the open ocean. Most tubenoses are truly pelagic, occurring well offshore near the continental shelf and beyond and rarely seen from land. Of nearly 100 species worldwide, about one third occur off the Northern California coast.

of their range, probably related to warmer water temperatures, with many fulmar corpses carried to beaches by current and tide. Most beach walkers mistake the carcasses for gulls, but the fulmar's "cracked" yellow bill with obvious tubes encompassing the nostrils reveals its true identity. Like other procellarids, the fulmar has a feather "pelt" with a rich, musty smell. Live fulmars occasionally gather near shore or on the beach, foraging on dead whales. At sea they prey on fish, squid, and zooplankton as well as offal. The fulmar's ocean flight is unique, alternating between gliding and shallow, stiff-winged flapping, low over the water, following the contours of the swells.

Northern Fulmars are polymorphic, and individuals may be "double-white," light, gray, mottled, or marbled; off our coast, gray birds usually outnumber all other forms by about nine to one. In the North Pacific, the darker color morph predominates in more southerly nesting colonies; lighter morphs predominate in more northerly colonies (Hatch and Nettleship 1998).

Sooty Shearwater — *Puffinus griseus*

LENGTH 15–17 IN.

These truly pelagic seabirds are aptly named because their remarkable flying ability allows them to shear close to the ocean's surface on fixed wings, then bank, flap, and shear low again, in an undulating flight path.

Sooty Shearwaters nest in the South Pacific on islands centered around New Zealand but spend the rest of their lives at sea. During the austral winter the entire population undertakes a remarkable journey along a route that, on a map, would resemble a tall figure eight, encircling both the north and south lobes of the greater Pacific Ocean. During this epic annual oceanic voyage, Sooties tend to travel in vast flocks, sometimes accompanied by much smaller numbers of other shearwater species, most commonly Buller's *(P. bulleri)* and Pink-footed *(P. creatopus)* (pl. 42). In the Northern Hemisphere, from late June into September, these "rivers of shearwaters," in numbers from 10,000 to over a million, veer very close to the Northern California shoreline, causing anyone who sees them (be they beach goers, surfers, or naturalists) to watch in awe, wondering "what in the world is going on?" Although there are six shearwater species that are regular and common off the Northern California shore (beyond sight from land), the Sooty is the most common and is included here because its nearshore presence often astonishes human observers. In fall, the Sooty Shearwater may be the most abundant bird species in California waters, especially in cool-water years.

Storm-Petrels

Family Hydrobatidae

The Latin family name, Hydrobatidae, means "walks on water" and accurately describes the distinctive foot-pattering behavior of these little sea sprites. All have webbing between the toes, which must aid their characteristic surface pattering behavior. The nickname "sea swallows" also gives a sense of their graceful, fluttering flight patterns. Of the seven North American species, three nest in our region and are covered here. All feed on small zooplankton and nekton, picking it off the sea surface with the delicate bill that has a pronounced hook on the upper mandible. Storm-petrels tend to forage at current confluences, fronts, and eddies where upwelling brings prey to the surface. Nesting sites are limited to offshore rocks without mammalian predators. All feed far from shore, visit their nest sites under the cover of darkness, and return to the same nest site year after year. Storm-petrels produce only one clutch of a single egg and raise only a single chick each year, and both sexes incubate and feed the young. All three species are monogamous, with both sexes sharing duties equally in nest attendance and care for nestlings. Like all Procellariiformes, and many other marine species, the storm-petrels are long-lived.

Fork-tailed Storm-Petrel *Oceanodroma furcata*

LENGTH 7.5–9 IN.; WING SPAN 18 IN.

Pl. 43

The breeding range of this pelagic beauty extends from the Aleutian Islands of Alaska southward to northernmost California—Del Norte and northern Humboldt Counties. This storm-petrel occurs strictly offshore, frequenting waters off the edge of the continental shelf, but may occasionally be blown in toward shore by strong west winds. During the nonbreeding season, Fork-tails may range farther south than their northerly nesting islands, into central California waters, as in Monterey Bay. Of the three species considered here, the Fork-tailed is the most likely to be found foraging on or around the floating carcasses of marine mammals.

> **During gale-force northwest winds in fall and winter, Leach's, Fork-tailed, and Ashy [Storm-Petrels] are sometimes blown to shore, and a proven place to see them up close at such times is from the U.S. Coast Guard jetty at Monterey harbor.**
> **—Rich Stallcup 1986, "Storm-Petrels"**

SPECIAL STATUS Bird Species of Special Concern, priority 3.

Leach's Storm-Petrel *Oceanodroma leucorhoa*

LENGTH 7–8 IN.; WING SPAN 17–19 IN.

This is "the most widespread Procellariiform breeding in the Northern Hemisphere" (Huntington et al. 1996). In the eastern Pacific the nesting range extends from northern Mexico (San Benito Island) to Alaska (Prince William Sound). In California, major nesting sites are at Castle Rock (Del Norte Co.), Trinidad Bay Rocks (Humboldt Co.), Little River Rock (Mendocino Co.), and Southeast Farallon Island (San Francisco Co.). Although quite common in Northern California waters, Leach's Storm-Petrel occurs well offshore, beyond the cold upwelling waters of the California Current. Leach's is seldom seen from shore, even near the nesting islands, because it visits its burrows and rock crevices only under the cloak of darkness.

Leach's comes in both white-rumped and dark-rumped forms, but the latter is more common south of our area.

Ashy Storm-Petrel ***Oceanodroma homochroa***

LENGTH 8 IN.; WING SPAN 18 IN.

The entire world breeding population of the Ashy Storm-Petrel is estimated to be on the order of 10,000 birds, fewer than many other species that are federally listed as "threatened" (Ainley 1995). Nesting colonies are restricted almost entirely to islands off the California coast; only one of the 17 known nesting localities is in Mexican waters. Perhaps half the population nests on the Southeast Farallon Islands off San Francisco. This limited distribution puts the species at risk. Populations of predatory gulls have increased dramatically in recent decades, and rodents (particularly the house mouse), predators against which storm-petrels have no defense, have invaded the larger islands where this storm-petrel nests, or did so formerly. Furthermore, after the nesting season, in fall, a large proportion of the population gathers en masse in Monterey Bay, placing the birds in jeopardy from oil spills or other catastrophic events.

When seen in flight at sea, the Ashy is characteristic of the tribe, swallow-like, feet dangling, sometimes pattering on the surface, "walking on water." Although this bird's plumage is entirely ashy gray, it is best identified by its flight pattern, raising its wings only to about the horizontal before each downstroke in a paddling motion; other storm-petrels in its range raise their wings much higher, slicing the air (Stallcup 1990).

SPECIAL STATUS Bird Species of Special Concern, priority 2; Bird of Conservation Concern in coastal California (USFWS 2008).

Cormorants and Pelicans

Family Phalacrocoracidae and Family Pelecanidae

These two closely related families are represented along the Northern California coast by three species of cormorants and two species of pelicans. All are adapted to an aquatic life and are rather large birds. Sizes range from the smallest Pelagic Cormorant *(Phalacrocorax pelagicus)*, with a wing span of 3.3 ft, to the American White Pelican *(Pelecanus erythrorhynchos)*, with a 9 ft wing span. Each species has all four toes webbed (totipalmate), is gregarious, and primarily eats fish. All have bare gular pouches (throat sacks). The Brandt's Cormorant *(Phalacrocorax penicillatus)*, Pelagic Cormorant, and Brown Pelican *(Pelecanus occidentalis)* are strictly coastal species, whereas the White Pelican and Double-crested Cormorant *(Phalacrocorax auritus)* also occur at inland waters. All three cormorants nest in sizable colonies along the Northern California coast, whereas Brown Pelicans nest south of our region and White Pelicans nest at saline lakes in the interior.

Brandt's Cormorant *Phalacrocorax penicillatus*

LENGTH 32–37 IN.; WING SPAN 42 IN.

Restricted to the Pacific coast of North America, from Alaska to Mexico, this species is closely associated with the cold upwelling waters of the California Current. The preponderance of the world population nests in Central and Northern California, with the largest colonies at Castle Rock (Del Norte Co.), Cape Vizcaino (Mendocino Co.), Southeast Farallon Island (San Francisco Co.), Bodega Rock (Sonoma Co.), and Bird Island (Monterey Co.). Numbers can vary dramatically from year to year, depending on seawater temperatures and abundance of prey. Overall, this is the most common cormorant all along the coast, sometimes numbering in the thousands at feeding frenzies, near shore or in larger bays, or hundreds perched on sea stacks. Following the nesting season, there is an apparent northward movement of both juveniles and adults (Ainley and Boekelheide 1990). Northward fall dispersal is also characteristic of several other California seabirds—the Brown Pelican, and Western and Heermann's Gulls.

These are highly colonial seabirds. Tens, hundreds, or even thousands of pairs may nest together on flat portions of large offshore sea stacks or islands. The remarkable courtship display—a slow-motion yoga-like pose exposing the brilliant cobalt-blue gular pouch to the sky with simultaneous wing fluttering and head ratcheting—is visible at headland nesting colonies from Monterey north to Del Norte Counties. With a close

A bird's foot that has all four toes connected by webbing is termed "totipalmate," a characteristic shared by several groups of North American waterbirds, including pelicans and cormorants, gannets and boobies, frigatebirds, anhingas, and tropicbirds. Swans, geese, and ducks have only three toes connected by webbing. *K.H.*

view, you may notice the white whiskers and plumes that decorate the adult breeding plumage, and the gemlike aquamarine eyes. Some adults steal nesting material from their neighbors, so nests are just beyond neck reach from all others.

Large flocks are often seen flying low over the ocean in long lines or V formations. The Brandt's tail is shorter than that of the other two species, and in flight, the nearly straight neck, lean profile, and obvious rounded head helps distinguish the Brandt's from the Double-crested Cormorant. The Brandt's bill is also noticeably thinner than that of the Double-crested.

Double-crested Cormorant *Phalacrocorax auritus*

LENGTH 31–36 IN.; WING SPAN UP TO 48 IN.

Pl. 44

The Double-crested Cormorant is *the* inland species in North America north of Mexico but also occurs along the coast, sometimes mixing in feeding flocks with Brandt's and even Pelagic Cormorants. Double-cresteds are more likely than the others to be found inside estuaries, or on reservoirs, ponds, or rivers. Flocks often fly high (sometimes in V formation in flocks of up to 200 individuals, reminiscent of geese) and over land. In flight, they have a long, rectangular head and a "kink" (Adam's apple) in the neck. All Double-cresteds have a yellow or orange gular pouch, and in their first year of life, many young are white ventrally. Indeed, the light underside of Double-cresteds in juvenal plumage is a reliable way to distinguish them from young Brandt's Cormorants. The "double-crest" appears in adults on the top of the head early in the nesting season, but is apparent for only a few months.

Cormorants, especially Double-cresteds, are often seen perched holding their wings spread, opened to the sun and air currents, to aid in drying the plumage. The cormorants' preen glands (the uropygial glands) apparently do not produce oils that provide the waterproofing common in other seabirds.

Nesting is opportunistic: on sea stacks, buoys, power poles, dead trees, and hunters' blinds.

Pelagic Cormorant *Phalacrocorax pelagicus*

LENGTH 25–30 IN.; WING SPAN 39 IN.

The smallest of the Pacific cormorants, Pelagics are daintier than our other two species. They occur all along the outer coast, especially along rocky shores and promontories, but are seldom gregarious (except when perched on headlands or sea stacks) and are usually far outnumbered by Brandt's Cormorants. Although named Pelagic, this species is rarely observed offshore, usually staying inshore where it forages among rocky habitats, even in the riptide or frothy intertidal waters. When prey is abundant, Pelagics may join foraging flocks of Brandt's Cormorant. In larger bays and estuaries, Pelagics tend to roost and forage near the mouth, rarely in the inner reaches. Rocky shoreline provides preferred foraging habitat where these agile divers hunt among the crevices for octopus, sculpin, eel, and other fare.

In flight, the neck is held out straight and the head is noticeably narrower than that of the other two species, and the tail is relatively long. The white flank patches are a great field mark *when they are present* but appear only from midwinter through spring, in adult alternate plumage.

Nesting is usually on predator-proof, narrow ledges of precipitous cliffs above the splash zone, but below the crest of the cliff. The simple structure is made of algae, sticks, and feathers cemented to the substrate with guano. Pelagics are not colonial, but limited availability of nest sites often places them in close proximity to neighbors of their own kind.

American White Pelican ***Pelecanus erythrorhynchos***

LENGTH 60–70 IN.; WING SPAN 114 IN.

Pl. 45

These magnificent beasts, with a wing span reaching over 9 ft, are rivaled in size only by the California Condor. "White pellies" nest in North America's interior, including California's Great Basin saline lakes, from March through July, typically occurring on the coast from July into early winter. Small squadrons of White Pelicans may appear around southern San Francisco Bay and Suisun Bay and on the outer coast at Abbott's Lagoon, Limantour and Drakes Esteros (Marin Co.), or Bodega Harbor (Sonoma Co.). In recent years these patterns have been changing, with earlier arrivals and later departures from the coast, an indication of nesting failures in the Great Basin due to drought, predation, and habitat loss. There have been no records of coastal nesting to date.

> **Often flocks will band together and, by beating their wings, drive a school of fishes into the shallows, where they gather up large numbers at every scoop of their big bag.**
> —Mrs. Eckstrom 1910, "The Bird Book," in *Birds of California*, edited by Irene Grosvenor Wheelock

Few avian sights are as enthralling as a flock of several dozen white pellies soaring on thermals high overhead against a cerulean California sky, wheeling slowly in unison, disappearing then appearing again like prehistoric shape-shifters, silent on the wing.

While Brown Pelicans are plunge divers, White Pelicans typically forage cooperatively from the surface of shallow water, often in flock synchrony, dipping and scooping their ungainly bills, then lifting them up vertically to swallow the prey. Primarily opportunistic fish eaters, especially of small schooling species, they have also been seen scooping filamentous algae from shallow marsh habitats. Whether these algal mats held small, entangled fish is unknown.

SPECIAL STATUS Bird Species of Special Concern, priority 1, due to population declines in the western states.

California Brown Pelican *Pelecanus occidentalis*

LENGTH 55–65 IN.; WING SPAN 78 IN.

Pl. 6, 46

Entirely a marine species, the Brown Pelican is a common sight—whether seen in a choreographed chorus line skimming the breakers, or roosting comfortably on a pier or piling in a harbor or bay. This gregarious bird is the California coastline's emblematic ambassador for the enlightened policies that served to save a dying species. It was nearly extirpated from North America's west coast by the early 1960s, when DDT and other pesticides flowing into the marine environment became more concentrated as they moved up the food chain, causing hormone disruption and thus eggshell thinning in the big birds. Brown Pelicans have staged a magnificent comeback, thanks to protections enforced by the Endangered Species Act, and are once again abundant all along the Northern California coastline. Birds that nest in Southern California migrate northward at the end of the breeding season, arriving along our coastal waters from April into October. A few usually linger through winter, more in some years than others, while most are at nesting islands in Mexico and off the coast of Southern California.

For what, after all, is more adroit than the flight of a Pelican? With three or four leisurely strokes the bird acquires a momentum with which he can glide with incredible accuracy just above the surface of the water. Or if he is hunting at a higher level, the bird is able to check his momentum, to put on the brakes midair . . . and plunge with the speed of thought upon his finny prey.
—Leon Dawson 1923, *The Birds of California*

The Heermann's Gull will steal fish directly from a pelican's bill whenever the opportunity arises, a habit known as kleptoparasitism. The gulls are more likely to pilfer from adult pelicans because they are more apt to come up with a beak full of fish. *K.H.*

Despite a seemingly awkward anatomy, Brown Pelicans are consummate plunge divers, pirouetting over the waves, tucking their wings, and diving headlong for schooling fish from high in the air. Heermann's Gulls, which Leon Dawson tagged "the worst of the pickpockets," speed in to scavenge spilled fish or those stunned by the concussion of the dive. Brown Pelicans, Heermann's Gulls, and Elegant Terns are known as "the three amigos."

SPECIAL STATUS Formerly federally Endangered (1970); "delisted due to recovery" (USFWS 2009).

Herons, Egrets, Bitterns, and Ibis

Family Ardeidae

Some herons and egrets are large and flamboyant, a familiar sight at marshes, lakes, and esteros or sometimes seen hunting in grasslands along the coast. These graceful birds were not always so common. Great and Snowy Egrets were common until the 1880s but were so thoroughly hunted for the plume trade that by the 1920s they were virtually extinct. Through later protection by law, they once again add statuesque living beauty to California wetlands.

American Bittern *Botaurus lentiginosus*

LENGTH 28–32 IN.; WING SPAN 42 IN.

Pl. 47

The furtive and stealthy American Bittern is a denizen of freshwater tule marshes and is therefore uncommon along the immediate coast. Extensive draining of wetlands has caused the numbers of the formerly more abundant bitterns to crash, and now they are easy to find only at waterfowl refuges in the central valleys. While white egrets have largely recovered from depredations by human hunters, the American Bittern is not likely to fully recover from the pervasive modification of its habitat.

The vertically streaked pattern on a bittern's underparts blends cryptically into the background of brown marsh vegetation—especially when the bird's neck is extended and the bill is "sky-pointing." To further the illusion of invisibility, the bird may even sway gently in concert with the cattails moving in the breeze.

The strange vocalization—a resonant, hollow, gulping sound—has lent the bird the curious but descriptive nicknames of "thunder-pumper" and "mire drum."

Although rather rare and sporadic along the immediate coast, the American Bittern is not entirely unexpected at coastal cattail-bulrush marshes, especially in fall and early winter. Most of the more southerly counties in our region report at least a few individuals on the annual Christmas Bird Count.

Least Bittern *Ixobrychus exilis*

LENGTH 11–14 IN.; WING SPAN 16 IN.

This diminutive ardeid's rarity along the coast, its preference for densely vegetated marshes, and its furtive behavior make encountering this species in our area highly unlikely. That said, there are a handful of nesting records from several of our more southerly coastal counties: Monterey, Santa Cruz, San Mateo, San Francisco, and Marin.

SPECIAL STATUS Bird Species of Special Concern, priority 2.

Great Blue Heron *Ardea herodias*

LENGTH 48–55 IN.; WING SPAN 72 IN.

Pl. 48

The Great Blue is the largest North American heron, standing about 4 ft tall and with an impressive wing span reaching 6 ft. It is a resident year-round in our region and rather common, but usually only lone individuals are seen at marshes, lakes, bays, and sometimes fields all along the coast. As diverse as the habitats it visits, the Great Blue's animal prey is always swallowed whole and headfirst. Great Blues are masters of patience and will stand stone still for many minutes awaiting the glint of a small fish or wiggle of a pollywog. Those standing in fields (especially after heavy rains that flood out rodents) are hunting gophers *(Thomomys)* or meadow voles *(Microtus)*, even lizards and snakes. In Santa Cruz or Monterey County you may see a Great Blue balancing atop a rolling kelp canopy, well offshore, foraging for sculpin. Like other long-legged wading birds, Great Blue Herons nest in colonies, often in association with egrets, and usually in groves of tall trees near water or marshlands. The bulky, stick-bundle nests are often clustered near egret nests, although in fewer numbers and more widely distributed, at least in the thoroughly studied San Francisco Bay Area.

Great Egret *Ardea alba*

LENGTH 40–48 IN.; WING SPAN 56 IN.

Pl. 49

Common and easily recognized by its tall, graceful stature and immaculate white plumage, and with a wing span of nearly 5 ft, the Great Egret is almost as large as the Great Blue Heron. The two species have similar foraging behavior, but the egret is less patient and tends to be more active in pursuit of its prey. Great Egrets often walk slowly and deliberately through shallow water or marshy vegetation, stalking prey. At some tidal wetlands—Palo Alto Baylands, Tomales Bay, Humboldt Bay—Great Egrets have learned to hunt the edges of extreme winter high tides, where they take surprising numbers of salt-marsh rodents, aquatic shrews, and even small marsh birds, including rails. Though they are most often seen feeding solitarily, Great Egrets will gather into groups to feed where prey is abundant, lining up along the shore of a tidal slough or in a shallow pond. Opportunistic feeders, Great Egrets consume a diverse array of animal prey—fish, crustaceans, amphibians, reptiles, birds, and small mammals.

Within nesting colonies amid groves of trees near foraging sites, nests are usually quite tightly spaced, often within a yard or two of one another, with dozens or scores of nests in a single colony. Rookeries are traditional, occupied year after year. Great Egrets are seasonally monogamous, and the sexes look alike, but males are slightly larger than females. The advertising and courtship display—performed at the nest site—is an elaborate and ritualized series of stretches, wing preens, and bows. Once the pair bond is secure, the pair performs greeting ceremonies when the mate arrives at the nest. The chicks are ungainly at first but become more attractive as the downy feathering develops.

The plumes (aigrettes) that extend from the scapulars to beyond the tail in breeding (alternate) plumage were hunted for the plume trade in the late 19th and early 20th centuries. Because this practice threatened to decimate egret populations, the National Audubon Society and the American Ornithologists' Union hired one of the country's first game wardens, Guy Monroe Bradley, to protect the Florida nesting colonies. Bradley was eventually shot and killed by plume hunters, a crime that inspired passage of the Migratory Bird Treaty Act (1918). Subsequently, egrets and other victims of the plume trade mounted a dramatic continent-wide recovery. As a result, the Great Egret is the symbol of the National Audubon Society.

Snowy Egret ***Egretta thula***

LENGTH 24–28 IN.; WING SPAN 40 IN.

Pl. 50

Just half the size of the Great Egret, an adult Snowy has a black bill and black legs with yellow "slippers." (The Great Egret's bill is yellow-orange and its legs and feet are black.) Snowy Egrets actively hunt aquatic animals that they locate by sight and capture with a crisp jab of the stiletto-like bill carried on the deceptively long neck. Prey is not "speared" but is taken in a quick "bite," then swallowed whole. Snowies are active in pursuit of prey, often running about in shallow water, sometimes with spread wings to shade the hunting area and cut down on glare. The Snowy Egret has a more diverse repertoire of foraging behaviors than any other member of the family. One common hunting method involves "foot stirring," in which the brightly colored golden foot is used to sift through the soft mud and dislodge worms, small fish, and crustaceans. (This is a foraging strategy also used, though less commonly, by several other waders, most notably yellowlegs and, on occasion, Black-belied Plovers.)

We sometimes see Snowy Egrets and Red-breasted Mergansers in a cooperative effort to herd small schooling fish. Working in tandem, swimming mergansers drive a school of anchovies toward the shallow shore where the egrets can reach them, then most of the school is pushed by egrets to slightly deeper water where they are easy prey for the ducks.

As in the Great Egret, filamentous feathers develop in the definitive breeding (alternate) plumage of Snowy Egrets, emerging as wispy plumes (aigrettes) from the crown, neck, and scapulars (shoulder feathers).

Cattle Egret *Bubulcus ibis*

LENGTH 30 IN.; WING SPAN 46 IN.

The Cattle Egret is a relatively recent arrival to coastal counties (1960s), and occurrence is sporadic and unpredictable. An initial increase in records through the 1980s, especially during fall and early winter, has subsided in recent years. Cattle Egrets seem to be most expected in late fall, when winter rains have moistened coastal pasturelands. They are most likely to be seen in damp pastures and lowlands, often where cows are browsing.

Green Heron *Butorides virescens*

LENGTH 17–20 IN.; WING SPAN 26 IN.

Pl. 51

Much less trusting than the Great Blue or the egrets, Green Herons are at home in riparian forests along slow-moving streams or around ponds and lakes. Bodies of water with overhanging willows are particularly attractive habitats for these shy stalkers. While a few are permanent residents, most disperse southward after the nesting season, even into Central America, returning to the California coast in the latter half of March. They are not colonial nesters in our region; the fragile-looking stick nest is usually 20 to 40 ft high in a willow or alder. When flushed, this little heron will often fly off, low over the water with its brushy crest erect, uttering a loud squawk and issuing a stream of white excrement, hence the nicknames "shitpoke" and "fly-up-the-creek."

The Green Heron is known to use bait to "fish"—dropping a feather or a willow leaf or even an insect on a creek surface to attract small fish, then successfully repeating the luring behavior.

Black-crowned Night-Heron *Nycticorax nycticorax*

LENGTH 26–30 IN.; WING SPAN 44 IN.

Pl. 52

Truly nocturnal, night-herons forage at night and rest during the day at communal roosts in dense shrubbery, usually near water. At dusk, they head out, flying in different directions to favorite foraging spots. The loud, hoarse "woc" calls emitted at dusk flight are among the most emblematic sounds of summer along the coast.

How often, in the gathering dusk of evening, have we heard its loud, choking squawk and, looking up, have seen its stocky form, dimly outlined against the gray sky and propelled by steady wing beats, as it wings its way high in the air towards its evening feeding place in some distant pond or marsh!
—A.C. Bent 1926, *Life Histories of North American Marsh Birds*

At breeding colonies, numerous pairs build nest bundles near each other in thick brush. Islands such as Alcatraz and Little Marin, the one in El Estero near Monterey Harbor, and Indian Island along the Samoa Blvd Bridge over Humboldt Bay are favored nesting sites.

Sometimes, when the weather is drizzly and gray, Black-crowned Night-Herons feed on into morning, but no matter how content they may be, they always have a grumpy expression.

White-faced Ibis ***Plegadis chihi***

LENGTH 18–22 IN.; WING SPAN 36 IN.

Another rarity along the coast, White-faced Ibis have been sighted increasingly in recent years, concurrent with their northward range expansion in California's Central Valley (Shuford et al. 1996). They are most likely to be encountered at a wetland edge in summer or fall, either singly or in a flock of up to several dozen birds. In flight, the heron-like shape, dark plumage, and long down-curved bill of the White-faced Ibis make identification fairly straightforward.

Vultures, Ospreys, Kites, Eagles, and Hawks

Family Cathartidae

Of three North American species in the vulture and condor family, one is ubiquitous in our region, one extirpated, and one absent. All are large, soaring scavengers, adapted to finding decomposing flesh over wide areas. Each has an unfeathered head and neck, large nasal openings, and broad, sail-like wings designed for effortless soaring. The large surface area of the wings relative to body mass is called "light wing loading." The talons are weak and these birds are unable to carry prey. Plumages are alike in male and female, though females average slightly larger. Although often lumped with the diurnal raptors—hawks, eagles, kites—vultures arose from a separate evolutionary line and, according to recent genetic studies, are more closely related to storks (family Ciconidae).

Vultures often roost communally, but they nest in solitude. The Turkey Vulture *(Cathartes aura)*, the only common family member in our area, is particularly adept at finding offal by smell, an adaptation that may account for its apparent success. Smell, an ability that is aided by a large olfactory bulb, is not as well developed in most birds, but an especially keen sense of smell is shared by the vultures and the tubenoses.

Turkey Vulture ***Cathartes aura***

LENGTH 27–30 IN.; WING SPAN 67 IN.

Pl. 53

This "janitor of the wild" needs no introduction to most Californians. Unless a storm front is in full swing, you can usually see a vulture or two soaring over the coastal plain on fixed wings, with wing tips held higher than the body, forming the letter *V* ("*V* is for vulture"), slowly tipping from side to side, "tippy gliding." While flying, "TVs" search for dead animals or for other circling vultures indicating the location of some deceased animal. Unlike most landlocked birds, the vulture's sense of smell is highly developed, facilitating its lifelong quest for decaying flesh. Its distinctive silver wing linings may function as visual signals to others of its kind, dispersed widely across the skyscape. Generous by disposition, nature's most thorough custodians gather in groups at a found carcass, but generally only one or two birds feed at a time. Turkey Vultures consume carrion before it rots and spoils. A beached dead seal or sea lion could keep a group of vultures fed for weeks. (The Turkey Vulture's tracks in the sand, with the dragging hind toe, are much like those of another common beach scavenger, the Raven, but larger.) Roadkill is consumed promptly by vultures; indeed, it is surprising that vultures do not fall victim more often to California's most efficient predator, the automobile.

Sociable birds, vultures often gather to roost in large flocks. After a damp night, they may assemble along fencerows or bare branches, holding wings akimbo to dry in the morning sunlight. So modest are they in their breeding efforts, vulture nests are rarely found. Their simple nests may be placed on cliff ledges, amid boulders, or even on the ground in hollowed-out trees. Two chicks are usually produced, wearing a surprising downy white plumage for the first several weeks of life.

California Condor ***Gymnogyps californianus***

LENGTH 46 IN.; WING SPAN 114 IN.

> *The Condor is the monarch of the air. We cannot say that there are not swifter birds or more agile birds, but there are none among the land birds . . . who have achieved a mastery [of flight] more unquestioned.*

So wrote William Leon Dawson in 1923, when condors were still flying free over California, although only in the most remote backcountry. In pre-European times, these sky masters ranged throughout the Pacific states, visiting whale carcasses on the seashore and salmon runs along our larger rivers, but in the 19th century the condor began to disappear from county after county, a downward slide that continued through the early 20th century. Shootings by humans started the decline, and lead poisoning and other environmental contaminants continued to feed the irreversible trend. Like the Grizzly Bear, the California Condor was just too large to coexist with the burgeoning humanity of the Golden State. The last two dozen or so free-flying individuals were captured in 1985 to 1987 and placed in captivity to initiate a captive-breeding program to save the species. After a decade of foster parenting by the Condor Recovery Team (in a cooperative effort by the San Diego Zoo, U.S. Fish and Wildlife Service, Ventana Wildlife Society, and others), these refugees from freedom were released into the wild. At this writing, the free-flying condors are surviving along the Big Sur coast, and several have bred in the wild. Recently some condors were seen feeding on the carcass of a gray whale on the Monterey shore, so it seems that there is hope, though slim, that these magnificent masters of the sky will reinhabit the California coast.

SPECIAL STATUS State Endangered; federally Endangered.

Family Pandionidae

Although sometimes placed in the same family as the hawks and eagles, the Osprey is now considered the sole member of the family Pandionidae, an indication of its specialized morphology. In the words of Roger Tory Peterson, the Osprey is a "world citizen," its range extending to all continents except

Antarctica. Its Greek name, *haliaetus*, translates to "sea eagle," an apt description of its "GISS," though not taxonomically accurate.

Most unusual are it's feet: The talons are extremely long and scimitar sharp; the outer toe rotates, allowing the bird to grasp its fish with two toes forward and two toes back. The base of the foot is covered with short spines (spicules), an adaptation for grasping slippery prey.

Ospreys are long-distance migrants, nesting in Northern California but wintering in Baja California and points south.

Osprey ***Pandion haliaetus***

LENGTH 24–27 IN.; WING SPAN 60–72 IN.

Pl. 54

Now common during spring and summer along much of the north coast, only a half century ago Ospreys were rare and in danger of extirpation. Like Brown Pelicans, Ospreys are "obligate piscivores," meaning they eat fish exclusively. Prior to the 1970s, agricultural pesticides, such as DDT, were used indiscriminately in California, contaminating the water and becoming concentrated in the food chain. Top predators, such as pelicans and Ospreys, experienced reproductive failure, and populations declined precipitously. These organochlorine contaminants were banned by the Environmental Protection Agency in 1972; the waterways and the fisheries became healthier, as did the fish eaters—a monumental conservation achievement.

Because of its large size, the Osprey may be confused with an eagle, but the flight is distinctive, with relatively long, strongly cambered wings and a pronounced bend at the "wrist." Wing beats are deep, interspersed with long glides. Ospreys often soar on fixed wings over water, then hover, tuck the wings in a headlong dive, and right themselves while extending talons as they reach the surface of the water. Foraging success varies with the species of fish taken, of course, but overall is relatively high as compared with most raptors. A study in Humboldt Bay (Poole et al. 2002) recorded 82 percent of 639 hunting efforts successful—56 percent on the first dive!

Most often seen around large estuaries and along large rivers on the coast, Ospreys are "semicolonial" breeders; that is, the nest sites are clustered in loose aggregations, usually along the banks of large rivers or the shores of estuaries, reservoirs, or lakes. A very vocal bird of prey, the Osprey's distinctive whistle is most often heard near the nest site. The large stick nests are conspicuous atop tall tees and snags, power towers, telephone poles, and even channel markers and duck blinds. Most Ospreys that nest in Northern California migrate south to spend winter on the coasts of Baja or western Mexico and south to Panama; some may go as far as South America. However, small numbers, mostly adult males, remain in the region over winter. The young do not return to the north coast until they are mature enough to breed at three or four years of age. Ospreys are so obliged to eat fish that they pack their lunch—sometimes carrying a fish carcass (or part of one) with them when they migrate, especially if traveling over dry ground.

Family Accipitridae

Kites, eagles, and hawks are diurnal birds of prey that have stout hooked beaks, strong talons for grasping prey, and large heads and eyes. Their vision is superb, akin to what humans see through a pair of high-end 8 power binoculars, thus the phrase "eyes like a hawk." Most hunt primarily by sight, but some have acute hearing as well (see Northern Harrier, White-tailed Kite). These raptors, once considered close relatives of the falcons, were recently determined to have a different ancestral lineage. The Accipitridae tend to have broader wings, bulkier bodies, and shorter, broader tails than the Falconidae. Many are darker above, lighter below, but there is great variation in coloration both among and within species. The larger members are sexually dimorphic, with females 10 to 20 percent larger than males, on average.

Recognizing the variation in shape is the first task in raptor identification. The soaring, open-country species have long, broad wings and relatively short, broad tails, adaptations for "riding thermals" and covering large tracts of open terrain. Eagles and Red-tailed Hawks are the prototypical soarers, so similar in shape that they are often confused in the field. (Size is difficult to judge when the soaring bird is silhouetted against a vast expanse of sky.) Several large species share this profile and are collectively called "the buteos," in reference to the Red-tail's genus. One of the buteos, the Red-shouldered Hawk, is similar in shape to others but is somewhat smaller, with a longer tail, designed more for a perch-and-pounce rather than a soaring strategy of hunting. Those species that forage through forest, shrub, and heavily vegetated habitats—"the accipiters"—have shorter broad wings and longer tails than the typical buteos, affording them more agility when chasing prey through an obstacle course of habitat. Kites (only one species here), the most falcon-like of this family, have evolved a unique technique for the successful capture of voles scurrying under the thatch of grasses and sedges—hovering, then parachuting down on rather pointed wings. Owl-like Northern Harriers, long winged and long tailed, ply marsh and pasture habitats as the kites do, but they use a different method to exploit a wider prey base. They course low over the terrain to scare up whatever creature lurks there, be it snake or quail, vole or meadowlark.

White-tailed Kite *Elanus leucurus*

LENGTH 14–16 IN.; WING SPAN 36–39 IN.

Pl. 55

The immaculate white and soft-gray plumage of this elegant raptor makes it conspicuous when perched in an open field atop a short tree or bush. The large head, short beak, and black shoulder patches are distinctive. The sexes are similar in size and plumage, but females are slightly darker on the back. The young have a diffuse rufous wash across the breast that is retained into early winter of the first year of life.

The habit of hovering to search for mice on the ground below has earned this kite its alternate name, "angel hawk," and also probably accounts for the name *kite*. (*Elanus* is Latin for this word.) These birds are often found at fields or marshy places on the immediate coast, hunting their favored prey, the California Meadow Vole *(Microtus californicus)*. So dependent are kites on voles that the boom-or-bust population cycle of the small rodent governs the distribution and behavior of the kite. Where vole populations are dense, especially in fall and winter, dozens of kites may gather together at communal roosts and hunting grounds.

In the 1920s, naturalist Ralph Hoffman wrote in *Birds of the Pacific States* that "there are probably not more than fifty pairs left in California," and by the 1930s, extinction was predicted in the state. In the last several decades, while declines associated with conversion of grasslands have been noted in Southern California, in our region numbers have increased dramatically, and most north coastal counties support healthy populations. Now, one is likely to encounter a kite, or several, during a day spent in the field, especially in pastoral lands away from the urban centers. The White-tailed Kite's distribution has been expanding northward in recent years.

Bald Eagle ***Haliaeetus leucocephalus***

LENGTH 30–44 IN.; WING SPAN 72–90 IN.

The national symbol needs no introduction, at least in adult (definitive) plumage. However, this, the second largest of our raptors (the California Condor is larger), takes four to five years to acquire the familiar white head and tail of a full adult. Immature birds can pose identification challenges, but these plumages are well covered in popular field guides. As in many other raptors, the female is noticeably larger (about 20 percent) than the male. The average adult male weighs about 9 pounds; the average female, about 11 pounds.

Bald Eagles have staged a dramatic increase in Northern California over the past decade, mirroring a continent-wide expansion since the early 1970s, yet another example of the positive result of wise conservation legislation (see the Osprey and Brown Pelican accounts). In recent years, Bald Eagles have become a fairly common sight at coastal embayments, large reservoirs, and the mouths of the larger rivers. They are often noticed in pursuit of Ospreys, attempting to pirate fish from their smaller cousins. During winter, with some regularity, Bald Eagles visit coastal lagoons, bays, and river mouths where dabbling ducks gather in numbers—from Bolinas northward. Nesting has been documented in all coastal counties from San Mateo County northward.

SPECIAL STATUS State Endangered; federally delisted; Bird of Conservation Concern in coastal California (USFWS 2008).

Northern Harrier *Circus cyaneus*

LENGTH 17–19 IN.; WING SPAN 42–54 IN.

Pl. 56

Coursing low and buoyantly over marshland or grassland, the harrier is ready to quarter and capture anything that moves—especially small mammals and birds. The former common name, Marsh Hawk, well describes its habitat preference. Northern Harriers are the most owl-like of our raptors, with facial disks that aid their acute hearing. Primarily "vole specialists," harriers also frequently take other small mammals, birds (meadowlarks, quail), and snakes.

The dihedral and rocking flight is similar to that of the Turkey Vulture, but the gray-and-white (adult male) or brown-and-tan (female and young) bodies with long tails and a flashing white rump will quickly dispel confusion. The Northern Harrier is the most "sexually color dimorphic" of the American raptors, meaning it is easy to tell male and female adults apart. (Subadults of both sexes are similar to adult females but darker brown above and russet below; there is no apparent difference between males and females in this formative plumage.) The pale-plumaged adult male Northern Harrier may be confused with the White-tailed Kite, as they both occupy the same habitats, but the kite is smaller and more slender and shorter tailed, tends to hover, and has black chevrons on its "shoulders."

Northern Harriers nest at favored places on the north coast but are more widespread in the nonbreeding season. The nest is placed on the ground, sometimes in a swale so densely vegetated with sedges or rushes that it limits intrusion by mammalian predators. In spring, the males perform a dramatic courtship flight display, wheeling in an undulating roller-coaster flight in the vicinity of the nest site. Once the pair bond is established and the female is attending the nest, the male delivers prey and the pair exchanges the offering in the air, displaying the aerobatic skills of this unique predator.

It seems that in winter, most adult females vacate the north coast while adult males remain through the year.

SPECIAL STATUS Bird Species of Special Concern, priority 3.

Sharp-shinned Hawk ***Accipiter striatus***

LENGTH 10–14 IN.; WING SPAN 21–26 IN.

Cooper's Hawk ***Accipiter cooperii***

LENGTH 15–18 IN.; WING SPAN 27–37 IN.

Pl. 57

These two accipiters are fairly common along the coast, smaller and larger examples, respectively, of the same basic model. Even experienced observers struggle with separating the two. Long tailed and short winged, the accipiters are designed for swift pursuit of their prey—usually smaller birds—through tangled vegetation and forest. Across open terrain, the "accips" have a characteristic flap-and-glide flight pattern, a few quick wing strokes followed by a long glide on fixed wings. These are among the most common raptors counted at coastal hawk migration sites. For example, over the 10 autumnal hawk watch seasons from 2000 to 2009, Golden Gate Raptor Observatory averaged over 6,600 accipiter sightings per year, about two-thirds of which were identified as Sharpies, one-third as Coopers. Incidentally, nearly 90 percent of those were immature birds.

Sharpies are rather "high-strung" whereas Cooper's Hawks are somewhat calmer in demeanor, a character trait that may explain why Cooper's have become more adapted to human-dominated environments, now nesting in urban parks throughout the San Francisco Bay Area. Sharpies, on the other hand, seek out more remote and densely forested areas for nesting. Because these are primarily bird predators, either species may key on seed feeding stations in suburban neighborhoods where there are easy pickings.

Northern Goshawk ***Accipiter gentilis***

LENGTH 21.5–24 IN.; WING SPAN 38.5–45 IN.

Although in the same genus as Cooper's and Sharp-shinned Hawks, the Northern Goshawk elevates the accipiter persona to an entirely other level. Goshawks are birds of the deep forest, powerful and seemingly fearless predators able to take prey more than twice their weight. As their name suggests, they nest only in the northernmost reaches of Northern California (Siskiyou Mountains), although farther south in the interior

mountains. Goshawks are rare in the Coast Ranges of Del Norte and Humboldt Counties, though more common in winter than summer. Their distribution in our region is similar to that of the Ruffed Grouse *(Bonasa umbellus)* and the Northern Flying Squirrel *(Glaucomys sabrinus)*, both favored prey species of this highly skilled raptor. Forestry practices in California have reduced available habitat and put this species at risk.
SPECIAL STATUS Bird Species of Special Concern, priority 3.

Red-shouldered Hawk *Buteo lineatus*

LENGTH 17–24 IN.; WING SPAN 36–42 IN.
Pl. 58

The Red-shouldered Hawk is an elegant (the western subspecies is *B. l. elegans*) and noble raptor, easily observed. It often sits on exposed branches of large riparian trees near river mouths or on telephone wires next to roads. The adult's breast is washed bright rufous, as are the shoulders. The tail is boldly striped black and white, and the flight feathers are checkered, also black and white. Young birds have plumage that is much more subdued, rather dark and mottled brown and tan, and they may be confused with a similar-sized accipiter (e.g., Cooper's Hawk). While the Cooper's appears rather long and lean (pl. 57), Red-shoulders seem brawnier in the shoulders and larger headed. The flap-and-glide flight across open terrain is also accipiter-like, a function of relatively short wings (for a buteo, but longer than those of an accipiter) and long tail, adaptations for pursuit of prey. The diet is variable, depending on season and availability—small mammals, reptiles, amphibians, and occasionally birds. Red-shoulders usually hunt from a low perch, looking down and dropping onto prey. This is a vocal raptor, especially on the nesting territory. Their chanting chain of whistled calls, in long series of almost two per second, is quite unlike the Red-tailed Hawk's, which is a wheezy, descending "skree."

Because Red-shouldered Hawks are more adaptable than Red-tailed Hawks (the Red-shoulder's main competitor) to habitats dominated by humans, they are now more populous than ever. Like the White-tailed Kite, Red-shoulders have been expanding their range northward into southern Oregon in recent decades.

IRRUPTIVE HAWKS

Two big buteos—the Ferruginous Hawk *(Buteo regalis)* and Rough-legged Hawk *(B. lagopus)*—may be encountered in open spaces along the coast in fall and winter. These species are "irruptive," with numbers highly variable from year to year. They are most likely to be confused with Red-tails, but somewhat different "body language" and behavior may alert the keen observer that he or she is encountering a different raptor. Consult your field guide!

Red-tailed Hawk ***Buteo jamaicensis***

LENGTH 20–24 IN.; WING SPAN 42–54 IN.

Pl. 59

By far the most common large soaring hawk year-round along the north coast, "western" Red-tails have highly variable plumage (dark, light, rufous, and mottled), a fact that can pose identification challenges. Young ones in their first year have brown, barred tails and are rangier than adults, with longer wings and tail. This probably gives the inexperienced birds some extra help while learning skills of flight, like training wheels for human toddlers. The tail feathers (rectrices) of adult Red-tailed Hawks are actually orange, not red. All except the darkest ("chocolate") Red-tails have a dark crescent mark (patagial patch) on the leading edge of the underside of the wing, just in from the bend of the wing (the wrist), usually visible as the hawk soars overhead.

The preponderance of large hawks seen along the north coast are Red-tails. They usually soar in circles, with tails broadly fanned and with motionless wings spread so the flight feathers look like long fingers.

The oft-heard voice, a harsh screech, is emblematic of the California Coast Ranges, asserted as both a warning and a declaration of territorial authority.

Resident year-round (nonmigratory) along much of the north coast, and monogamous, Red-tails build nests in trees or power poles, and the same pair may use the same nest year after year. Pairs can be seen flying and roosting in tandem even in midwinter. Red-tails are bold and aggressive in defense of nest sites and territory, rarely intimidated even by eagles or Great Horned Owls. Like most hawks, Red-tails are opportunistic hunters, but medium-sized mammals—brush rabbits and ground squirrels—seem to be important prey items.

Golden Eagle ***Aquila chrysaetos***

LENGTH 28–33 IN.; WING SPAN 78–90 IN.

Pl. 60

Primarily birds of the open country, especially the inner Coast Ranges, Golden Eagles nest in the coastal counties in limited numbers, and young birds occasionally patrol the coastal plain, especially during fall. Accomplished predators, Goldens will take any prey that is available but seem to specialize in jackrabbits and ground squirrels. Separating immature Golden and Bald Eagles is tricky; the color pattern of the underside of the wing, the tail shape, the bill size, and overall proportions are helpful clues. On birds in flight, pay special attention to the pattern of the light-colored feathering on the underside of the wing.

Rails, Gallinules, and Coots

Family Rallidae

Rails live in the shadowy tangles of marsh plants—tules, cattails, sedges, and pickleweed. Their bodies are laterally compressed, so they can slip through narrow passageways. The phrase "skinny as a rail" is derived from this peculiar anatomy. Feathering is dappled—shades of brown, tan, olive, black, and gray, some speckling of white—and provides cryptic plumages. The solitary rails—Clapper, Virginia, Sora, Black, and Yellow—are shy and cautious, rarely venturing more than a quick scamper from cover. American Coot and Common Gallinule are less furtive and more sociable than the solitary rails.

Yellow Rail *Coturnicops noveboracensis*

LENGTH 7–8 IN.; WING SPAN 11 IN.

This and the Black Rail are among the most furtive of North American birds. The Yellow Rail breeds in interior freshwater wetlands, many of which are seasonal, and small numbers arrive at the coast to winter in brackish and tidal salt marshes. San Francisco Bay and associated outer coastal marshes are the only areas where Yellow Rails are known to occur regularly on the coast. In winter they are silent, so the only chance to see one is when high tides or flood waters force them out of the densely vegetated marsh habitats.

SPECIAL STATUS Bird Species of Special Concern, priority 2.

"California" Black Rail *Laterallus jamaicensis coturniculus*

LENGTH 6 IN.; WING SPAN 9 IN.

Pl. 62

The Black Rail is a furtive denizen of tidal marshes, and occasionally brackish or fresh marshes, in the San Francisco Bay Area. San Pablo and Suisun Bays support the preponderance of the world population of the California subspecies. Black Rails are resident in the high marsh zone, requiring damp soils and a thick cover of native wetland vegetation. Although rarely seen, the Black Rail can be heard calling, advertising territory, during spring and summer in appropriate habitats. The call is a uniquely modulated three-note "kik-kek-grr," sometimes repeated incessantly at dawn, at dusk, or even on moonlit nights. Recent research has found that Black Rails occupy very small territories within tidal marsh habitats, therefore densities can be quite high where extensive marshlands exist. Because of their diminutive size, when forced out of their preferred habitat by extreme high tides or floods, they may be preyed upon by egrets and herons, and sometimes by gulls and hawks.

SPECIAL STATUS State Threatened.

The Black Rail flies only when conditions require it to leave the marsh and seek refuge on higher ground. Unfortunately, Great Egrets cue on flood tides and pursue any refugee small enough to capture and consume, be it crab, vole, or bird. The size advantage of the egret is difficult to overcome, no matter how rapid the rail's wing beats. *K.H.*

"California" Clapper Rail ***Rallus longirostris obsoletus***

LENGTH 13–15 IN.; WING SPAN 18 IN.

Pl. 63

Formerly more widespread at lagoons and estuaries along the outer coast, the distribution of the "California" Clapper Rail is now restricted to only the remaining tidal marshes of San Francisco Bay. (Two other subspecies occur, only in Southern California.) This is a large rail closely associated with channels and sloughs in larger, fully tidal marshes. Present year-round, Clappers initiate breeding very early in spring, timing their nesting to avoid the extreme tides that inundate their chosen habitat twice a year—in midwinter and around the summer solstice.

Like most other rails, Clappers are rather secretive, remaining hidden in the sloughs and salt marshes, rarely flying or venturing out into the open. You may hear the boisterous "clattering" calls emanating from marshes along the bay shore, usually at dawn or dusk. These are the advertising calls of duetting pairs, exclaiming their territories to their neighbors. During the extreme diurnal high tides of winter, rails are forced out of the lower marsh, taking refuge in adjoining habitat. A good place to see Clapper Rails is during winter high tides at the Palo Alto Baylands Preserve at the east end of Embarcadero Road in East Palo Alto.

At this writing, taxonomists are revaluating the relationships among various populations of Clapper Rails, and it is expected that in the future the western subspecies may be split from the eastern and South American subspecies. The proposed new name for the western group is *Rallus obsoletus.*

SPECIAL STATUS Federally Endangered; State Endangered.

Virginia Rail *Rallus limicola*

LENGTH 9–10 IN.; WING SPAN 13 IN.

Pl. 64

Virginia Rails are present year-round in fresh and brackish coastal marshes, but numbers swell with an influx of wintering birds. Although furtive, Virginias are occasionally seen at the marsh edge. This medium-sized marsh bird is generally nocturnal or crepuscular and when seen during the day is never more than a quick dash from the cover of marsh vegetation. The bill is brightly colored and curved, an adaptation for probing in the marsh mud for small aquatic fauna—beetles, snails, larvae, crustaceans, and whatever else is found in these swampy digs.

More often heard than seen, the Virginia Rail's most frequent call is a descending emphysemic cackling "grunt," often given in duet. This call is contagious; where densities are high, you may hear outbursts of grunts throughout the marsh. Less common are the "tick-it" (or "ki-dick") of the "kicker" call, and an emphatic squeeky "keeu." That last call is easily confused with the calmer "weep" call of the Sora *(Porzana carolina)*, a species that often occurs in the same habitat. On the coast the Virginia Rail tends to outnumber the Sora, considerably.

Sora *Porzana carolina*

LENGTH 8–9 IN.; WING SPAN 14 IN.

With habits similar to those of Virginia Rails, the plumper Soras may sometimes be glimpsed in shadows at the muddy edges of marsh vegetation, preferring freshwater habitats. On the coast, Soras are not nearly as common as Virginias. The Sora has a stubby yellow bill more adapted to eating seeds than the probing bill of the Virginia Rail, which may explain why these two marsh dwellers are able to share habitat within the same marsh. Although the sexes are similar in plumage, the bill of the adult female Sora is not as bright yellow as the male's, and she has less black on the throat. The Sora is not likely to be confused with the smaller, rarer, and much more secretive Yellow Rail. The bird's name is an allusion to its frequent call: a loud, whistled, ascending "soorraaah." Also unique to the species is a rapid, squealing, descending whinny.

Common Gallinule *Gallinula galeata*

LENGTH 14 IN.; WING SPAN 20–21 IN.

One of the social rails, the Common Gallinule (formerly called Common Moorhen) is closely allied with the more familiar American Coot, both in looks and behavior. Never as common as coots, and more retiring, gallinules are regularly present only in the more southerly portions of our region. They are mostly found in freshwater habitats—ponds and lakes with dense fringing vegetation—rarely in more brackish environments. Like coots, gallinules are somewhat gregarious in winter, but quite territorial when nesting. An interesting aspect of the Common Gallinule's breeding behavior involves "juvenile helping," in which chicks from the first brood help raise their younger siblings from the second brood. This cooperative effort includes not only feeding, but also brooding, warning the chicks about predators, and defending territory.

When in full feather of breeding dress (males and females alike), Common Gallinules are unmistakable because of the bright yellow-and-red bill with a large frontal shield extending up between the eyes, as well as an obvious white horizontal stripe below the wing, along the waterline.

Gallinules seem to have increased, at least in the more

southerly counties of our region, over the past decade. This trend has also been apparent in several other western states (Sauer et al. 1996).

American Coot *Fulica americana*

LENGTH 15–16 IN.; WING SPAN 24 IN.

Pl. 61, 65

American Coot is the most abundant and widely distributed member of the rail family in North America. Unlike our other rails, coots are highly gregarious and fairly trusting, gathering in sizable groups to forage on algae along tidal flats or congregating on open water. Coots are quite wonderful and thoroughly underappreciated, as indicated by the layman's epithet, "mud hen." At a distance, a coot could be mistaken for a duck, but taxonomically, as a member of the family Rallidae, coots are more closely related to Sandhill Cranes than they are to Mallards. On closer inspection, the American Coot's distinctly shaped white bill, laterally flattened, and the smallish head reveal its true identity. Omnivores, coots will graze on lawns, "dabble" for aquatic vegetation, and eat seeds and bread, mollusks, and other invertebrates. American Coots make nervous noises when in flocks, one of which is a perfect imitation of the rubber bathtub duck. Their winter congeniality breaks down during the nesting season, however. Protective of their personal space on a nesting lake or pond, coots become more confrontational, challenging interlopers—head down, taillights up, motoring after rivals that blunder across invisible turf boundaries.

Nests are built in stands of dense emergent vegetation along the fringes of marshes, ponds, and lakes, close to or floating on the water's surface. Like other rails, coots have rather large clutches, holding a half dozen to even a dozen eggs, with older females laying a greater number. Second broods are common, but hold fewer eggs. For the first few days of life, in summer, the small chicks follow the parent around the nesting site, often venturing into open water. These downy young are charmers—tiny, with flaming crimson-bronze downy feathering around the head and a scarlet forehead and bill. This colorful down is lost by about two weeks of age (postnatal molt), and the mouse-gray juvenal plumage is acquired.

Shorebirds

Family Recurvirostridae

Avocets and stilts are supermodels among the shorebirds—slender, long legged and long necked, elegantly outfitted. The family has a cosmopolitan distribution, but only two species are in our region, both rather locally common. Sexes are similar, though males are slightly larger than females. Gregarious among their kind, they often gather to forage or roost in large flocks. Shallow water is the preferred habitat. They are a vociferous lot, especially around nesting territories, where interlopers are loudly scolded, circled in flight, and even strafed. Nests ("scrapes") are placed in the open, on islands or levees mostly inaccessible to mammalian predators. As with other shorebirds and ground-nesting species, chicks are precocious, leaving the scrape shortly after hatching, to follow the parents and learn the ways of the world.

Recurvirostridae, from the Latin, refers to the avocet's bill shape, literally meaning "turned back" or "upturned," an adaptation for scything through shallows for prey—brine flies, insect larvae, and small crustaceans. Stilts have a straighter, but equally narrow, bill and peck rather than sweep for prey.

Large wetland restoration efforts over the last several decades, especially in San Francisco Bay, have increased the availability of the habitat for these two species, and their populations have responded quickly, showing rather dramatic regional increases.

Black-necked Stilt ***Himantopus mexicanus***

LENGTH 14–15 IN.; WING SPAN 28 IN.

Pl. 71

The well-dressed and lanky Black-necked Stilt is rare on the outer coast except for north coastal Monterey County at Moss Landing's Elkhorn Slough, where there is a source population. The coastal breeding range extends north only to Marin County. At salt ponds and other alkali shallows, stilts are abundant along the shores of San Francisco Bay, where hundreds may be seen together. The largest concentrations are resident in salt evaporation ponds of Sonoma, Napa, and Solano Counties in San Pablo Bay and in Alameda, Santa Clara, and San Mateo Counties on the San Francisco Bay shoreline.

A preference for evaporation ponds and waters associated with the urbanized estuary is a double-edged sword: While extensive tracts of habitat are available, these ponds tend to concentrate contaminants in the food web and cause harmful effects on the reproductive success of stilts. Likewise, plans to restore much of the abandoned salt panne habitat in San Francisco Bay to tidal influence may reduce the habitat available for nesting stilts.

Stilts share similarities with avocets, and the two types of birds are often seen together. Primarily visual foragers, stilts tend to hunt more by pecking than scything. They are extremely vocal and territorial in the vicinity of the nest or chicks, performing wild histrionics when perceived predators approach.

American Avocet *Recurvirostra americana*

LENGTH 17–19 IN.; WING SPAN 31 IN.

Pl. 3, 72

This is an elegant, long-legged, and striking shorebird common or abundant in the southern section of our region, rather rare north of Sonoma County except at Humboldt Bay. Many, or most, of the avocets that live at wetlands around San Francisco Bay and Elkhorn Slough (Moss Landing) are year-round residents, nesting in the salt flats and on islands and levees. The avocet mating system is known as "colonial monogamy," meaning the males and females form pairs, then gather into loose nesting colonies with sexes sharing parental duties. Those that winter at some estuaries on the outer coast (Humboldt Bay, Bodega Harbor, and Bolinas Lagoon) are migratory, leaving in early spring toward their inland nesting areas, perhaps in eastern Oregon, Idaho, or Alberta. Wintering roosting flocks can be quite large, which is thought to allow a relaxation of vigilance (Robinson et al. 1997).

The upturned (recurved) bill of the avocet is often used for "scything" aquatic insects from the shallow water, but feeding methods vary and may include pecking, plunging, and snatching. Male avocets have somewhat straighter bills than do females, perhaps so they can harvest slightly different prey from the same waters during the chick-rearing period. This evolutionary strategy, called resource partitioning, is not uncommon in birds.

Unlike the stilt, the avocet is a rather good swimmer, and small foraging flocks are sometimes seen paddling through water deeper than their leg length. Newly fledged chicks, too, can swim almost immediately and even dive to avoid predation. In our area, aerial predators—Peregrine Falcon, Northern Harrier, Barn Owl—probably present the greatest threat; undoubtedly gulls and ravens take eggs and chicks. But anyone who has approached an avocet nesting area has experienced the vehement protective actions of the adults—frantic swooping and strident calling—attempting to distract and discourage the trespasser.

Family Charadriidae

Plovers belong to a morphologically distinct group of shorebirds—rather stout with large eyes, short necks, relatively short bills, and long, pointed wings. Most are long-distance travelers, traversing continents, even oceans, in their annual migrations. Plovers hunt by sight on mudflats or beaches, or short-grass pasture and prairie. They are not probers, drillers, or stitchers; their method of foraging is "run, stop, listen, look, pick," feeding like a robin on a lawn. Plovers pick invertebrates off the surface of the ground, or occasionally pull prey from shallow holes, but with their stubby bills, they are unable to probe. With their relatively large eyes, they are able to forage in very low-light conditions, even at night.

The difference in plumage is slight between sexes as well as between juveniles and adults. Difference in size between sexes is also minimal, not discernable in the field. Plovers gather in smaller, less compact flocks than do the sandpipers, each individual maintaining a comfortable degree of personal space. Most plovers, especially the larger species, are very vocal and easily flushed when approached.

Black-bellied Plover ***Pluvialis squatarola***

LENGTH 11–12 IN.; WING SPAN 28–32 IN.

Pl. 66

Rather large and gregarious shorebirds, Black-bellied Plovers often fly and roost in sizable flocks, but they spread out once they leave the flock to feed on tidal flats. Because they visually search for surface or near-surface prey, each one needs room to move around. Black-bellies also forage and roost on some outer sand beaches, especially when tidal flats in nearby estuaries are flooded.

Named for a stunning breeding (alternative) plumage, worn here only in April before the birds leave for their high Arctic breeding grounds, the Black-bellied Plover has winter (basic) plumage that is much subdued, grayish brown above, white below. In Europe the same species is known as the Gray Plover, a name more descriptive of the muted plumage worn in winter. From August through March, this is the most common plover along our coast. It is a visual forager with a large eye, an attribute that also provides it with an acute alertness to predators. One of the most recognizable and emblematic sounds of tidal flats in winter is the three-note, whistled "pee-oo-wee" flight call of the Black-bellied Plover. Some arrivals in late July retain some of their spiffy breeding dress; likewise, some birds departing in April are already acquiring theirs. In flight the black "armpits" (axillars) help distinguish this from the similar-sized, and much less common, Golden-Plover.

"Western" Snowy Plover ***Charadrius nivosus nivosus***

LENGTH 6–7 IN.; WING SPAN 17 IN.

Pl. 67

The Western Snowy Plover is a compact little bird of the outer beaches and bay salt flats. The preference for nesting on the coastal beaches—a habitat habituated by humans and ravens—has imposed severe pressures on the California population, now decidedly "at risk." During summer most Snowy Plovers nesting on the outer coast are encircled by predator exclosures, put in place by biologists and regulatory agencies. Rope fences with occasional Snowy Plover interpretive signs cordon off nesting territories against humans and their dogs. Those Snowy Plovers that nest on salt pannes and pond margins around San Francisco Bay (where about 10 percent of the population breeds) may fare better than the outer-coast individuals. One study found that plovers nesting within about 330 ft of Least Tern colonies had higher nesting success than those nesting farther from tern colonies (Powell 2001). The Snowy Plover—as well as the Eurasian counterpart, the Kentish Plover—has an unusual mating system in which the female, and less commonly the male, abandons the first brood shortly after it hatches, to nest again but with a different mate.

Even though overall numbers are low, Snowies are not hard to find at broad sandy beaches along the coast, especially in winter when they gather in flocks. They rarely frequent the wet sand of the lower beach, more frequently roosting and foraging in the higher, drier sand. When winds buffet the coast, Snowies often crouch for refuge in footprints, in deflation pockets formed by the wind, or behind shore-cast debris or driftwood.

Any of the wider outer beaches with extensive dune systems that parallel the shore may be visited by Snowy Plovers. Even Ocean Beach in San Francisco, with heavy dog and human use, harbors a substantial winter flock.

SPECIAL STATUS Federally Threatened.

Semipalmated Plover ***Charadrius semipalmatus***

LENGTH 7–8 IN.

Pl. 68

Semipalmated Plovers look much like the more familiar Killdeer but are smaller and have only one black neck ring instead of two. Also, the complete neck ring, darker dorsum, and orange legs will distinguish "semipalms" from the paler, and slightly smaller, Snowy Plovers. Semis and Snowies often roost together above the wrack line on ocean beaches. Although never abundant and with a tendency to occur in smaller flocks than many other shorebirds, a few dozen Semis may be counted among the other shorebirds on the beach or tidal flat during migratory pulses. Semis are highly coastal in winter, frequenting marine-influenced environments, rarely occurring inland. Like other plovers, they search for prey visually, scurrying across the flats to pick up small invertebrates. When foraging, individuals tend to spread out across the flats rather than cluster into tight groups as do many small sandpipers. The voice is distinctive, a clarion whistled, two-note "ker-wee," rising on the second syllable.

This is a bird of the far north—nesting in the Arctic and sub-Arctic—that spends the nonbreeding season along the Pacific coast from Northern California southward. It is also a long-distance migrant; this habit is enabled by long, slender wings allowing powerful flight.

Sexes are similar in appearance, but the black markings on the female may be duller or flecked with brown and the white stripe above the eyes more distinct. The bird's clumsy name comes from the fact that it has partial webbing (palmations) between all of its toes, a characteristic difficult to observe under normal viewing conditions. A similar species in Europe has webbing between only two toes on each foot.

Killdeer ***Charadrius vociferus***

LENGTH 10–11 IN.; WING SPAN 24 IN.

Pl. 69

Along with the Snowy Plover, oystercatcher, avocet, and stilt, the Killdeer is one of the few species of shorebird that nests in our region. Although a true shorebird, and quite common, the Killdeer's habitat preference is not for mudflat, marsh, beach, or rocky shore. Instead, Killdeers are found more frequently on lawns, plowed fields, barren grounds, human-altered habitats, and gravelly places. The nest is often placed among stones on the shoulders of roads, on creek banks, or even on roofs; the eggs are speckled cryptically to blend in with the cobbled substrate. Typically, four eggs comprise a clutch. When a potential predator (like you) goes near the nest site, one of the adults performs a "broken-wing act," feigning injury and calling stridently while floundering away from the nest. When the predator is safely detoured by this distraction display, the distressed performer regains composure and flies away, and returns to the nest site when the coast is clear.

The chicks are precocious when hatched, meaning they can leave the nest within hours, as soon as their downy feathers dry. Apparently the adults do not feed the chicks, but rather lead them to feeding areas and teach by example. The downy chick is nearly as cryptic as the eggs, with disruptive patterns of black and white and mottled brown tones on the back. The legs are comically long and the feet are huge, way out of proportion to the tiny ball of fluff that sits atop. The chicks molt and attain a juvenal plumage similar to the adult's within about three weeks of hatching.

Killdeers can be very noisy, and the specific name *vociferus* (i.e., vociferous) is appropriate. One of the common sounds around rural areas is the clamorous vocalization, the eponymous alarm call—"killdeeeer, killdeeeer, killdeeeer."

Family Haematopodidae

Black Oystercatcher ***Haematopus bachmani***

LENGTH 17–19 IN.; WING SPAN 32 IN.

Pl. 5, 70

Oystercatchers may be seen year-round roosting or foraging on intertidal rocky shore or sea stacks, habitats from which they rarely stray. The Black Oystercatcher is a stout shorebird, well adapted to life in the churning splash zone. The sharp call note is piercing and can be heard above the roar of breaking waves; it is especially noticeable when lengthened into a series of raucous trills. This vocalization usually indicates that a territorial dispute is in the works.

The Black Oystercatcher's black and dark-brown plumage matches dark, wet rocks, algae, and mussel beds so well that it is most often located when an observer spots its thick, pink legs and goes up from there! The tomato-red bill—straight, long, and heavy—is used as a tool to open mussels and pry other mollusks from rocks. The bill of the adult is all red; that of hatching-year birds is red with a dark tip. The yellow iris and red orbital ring give the bird a look at once comical and serious, quite an amazing countenance to behold.

Family Scolopacidae

This highly diverse family of sandpipers, phalaropes, and their allies encompasses a wide variety of structure, plumages, and behaviors. The relationship between the plovers (Charadriidae) and these sandpipers is unclear but provides a basis for comparison. This group is better designed for wading in shallow water, and its members have relatively smaller heads and eyes than the plovers but longer, thinner bills. The bill of the probing species is highly sensitive, bundled with nerves at the tip, an adaptation for tactile (rather than visual) foraging. This bill sensitivity allows the Scolopacidae to detect hidden and buried prey items and is used for various foraging techniques—discussed

Shorebirds are designed in a marvelous diversity of shapes and sizes. Pictured here, for scale, is a Long-billed Curlew (in part) with Least Sandpipers in the background scurrying past. *K.H.*

under each species account, below. To generalize (and there are exceptions to every generality), shorebirds that probe in softer substrates have the most curved bills, and species that probe (or pick) in firmer substrates have straighter bills (Piersma 1995). Sandpipers that forage in more saline environments have salt glands, located between the eyes, that concentrate salts for excretion by way of the nasal cavity. In those species that occur in freshwater habitats, salt glands are much reduced or absent (e.g., Wilson's Snipe).

Most of the sandpipers are long-distance migrants with long, pointy wings that aid in their often epic journeys from nesting grounds to wintering grounds. In the nonbreeding season, most sandpipers wear a rather dull plumage—pale below and somewhat dark above—whereas many take on a much brighter, more colorful breeding dress. As you will see from the species accounts below, the family embraces a wide array of breeding strategies, and sex and size dimorphism is equally diverse. Many of the shorebirds are highly gregarious in the nonbreeding season, migrating, foraging, and roosting in rather large flocks.

Spotted Sandpiper *Actitis macularius*

LENGTH 6–8 IN.; WING SPAN 15 IN.

Pl. 74

Spotted Sandpipers do not occur in flocks and when on the coast are usually found singly on the rocky shore, at creek mouths, along the shores of reservoirs and ponds, or on wharves, pilings, and even moored boats. The unique and stunning alternate (breeding) plumage—white, with large round black spots—is worn only from April to August, the period when they are at nesting places in the interior. The relatively plain basic (winter) plumage is similar to that of several other shorebirds, but the Spotted's small size and habitual "teetering" with the rear end of the body will aid in identification. (Other tringine sandpipers teeter also, but with the front of the body.) When flushed from the shore, the flight pattern—shallow, stiff wing beats, low over the water—is unique among the shorebirds and easily recognized. (Unlike many of their family members, Spotted Sandpipers have relatively short wings and are not particularly long-distance migrants.)

Widely distributed across North America, Spotted Sandpipers nest on sand or gravel bars along creeks and rivers throughout much of the United States and Canada. Unlike most other shorebirds, the Spotted Sandpipers' sex roles are reversed (classic polyandry): females initiate courtship (find territory, attract a mate), and males take on the majority of the parental responsibilities (incubating and brooding).

TRINGINE SANDPIPERS

Five species of sandpipers (tribe Tringini) in the genus *Tringa* share some broad characteristics that are helpful in recognizing this group in the field. All are relatively long legged (thus "tall"), have a bill at least as long as the width of the head in profile, have relatively large eyes (visual foragers), are active and especially boisterous, and are often quick to scold or sound an alarm when an intruder threatens.

Solitary Sandpiper ***Tringa solitaria***

LENGTH 8–9 IN.; WING SPAN 22 IN.

Aptly named, this tringine sandpiper is rarely encountered in our region. It is most likely to be seen bobbing in place along a quiet stream or pond during fall migration, seldom in tidal wetlands, and almost always solo. During spring two or more individuals may arrive at some shallow pond, sandbar, or creek shore, but the visit is usually fleeting.

Wandering Tattler ***Tringa incana***

LENGTH 10–12 IN.; WING SPAN 26 IN.

Tattlers are rocky shore birds (a small suite of species including oystercatchers, Surfbirds, Rock Sandpipers, and Black Turnstones) that specialize in foraging wave-battered, rocky shorelines off the outer coast. Although hypercoastal, occasionally Wandering Tattlers are found in larger embayments, along seawalls or rock jetties. Indeed, the jetties at Humboldt Bay,

Bodega Bay, and Princeton Harbor are some of the most accessible sites to find them. Southbound migrants, still in breeding plumage, pass along these shores as early as midsummer; wintering birds in basic plumage apparently arrive somewhat later. Turnstones and Surfbirds are usually gregarious, but Wandering Tattlers are mostly solitary, only joining the other "rockies" to roost on some tall rock or sea stack, above the splash zone, out of reach of the longest fingers of the tide.

The uniform gray back in winter (basic) plumage often blends in with the Wandering Tattler's inaccessible, slick-rock habitat, making this species easy to overlook. The loud flight call, a trilling "trea-trea-trea-tree," is audible even over crashing waves, a helpful clue to its presence.

Except when flying high and far during migration, Wandering Tattlers are seldom far from rocks. They nest in scree or boulder-studded tundra slopes on mountains in Alaska, far from the sea.

Greater Yellowlegs *Tringa melanoleuca*

LENGTH 14–15 IN.; WING SPAN 28 IN.

Pl. 75

Yellowlegs are slender sandpipers, well named. Both Greater and Lesser Yellowlegs have long, bright yellow legs. They are not generally gregarious and they occur in groups only at exceptionally productive forage opportunities or during migration. A noticeably long-legged and solitary shorebird walking along a slough's edge or marsh shoreline will often be a yellowlegs. If it has a rather long and (barely perceptible) upturned (recurved) bill, it is likely a Greater Yellowlegs.

Greater Yellowlegs is the more common of the two species, found at most coastal wetlands except in May and June, when it nests in Canadian and Alaskan spruce bogs. Freshwater and lightly brackish marshes are preferred over tide flats or coastal shore. When foraging, Greater Yellowlegs are often active, pursuing prey by sight through shallow water and lunging or stabbing. A member of the "tattler" group of shorebirds, the Greater Yellowlegs is very vocal, using its clear, clangorous alarm call, "tue-tue-tue . . . ," to scold interlopers.

A Willet and a Long-billed Curlew struggle over the possession of a ghost shrimp. The curlew's bill is designed to probe and extract the ghost shrimp from its deep burrow; the Willet's personality focuses on any opportunity to grab a meal. *K.H.*

Lesser Yellowlegs ***Tringa flavipes***

LENGTH 9–11 IN.; WING SPAN 24 IN.

This is a diminutive version of the Greater Yellowlegs, with very similar behavior and habitat, and both species may be seen together during fall migration. Lesser Yellowlegs are uncommon in our region, occurring mostly in fall (mid-July through September) on the coast, where nearly all transients are juveniles (young birds of the year). If the individual's bill is thin, stiletto straight, and no longer than the head in profile, it may be a Lesser. The Lesser usually pecks prey from the surface of the water or the shore, rarely running after prey as does its larger cousin. Although Lesser Yellowlegs are usually seen in small numbers during migration, larger flocks (100 plus) have been reported from the more northerly counties (e.g., Lake Talawa, Del Norte Co.). Many more Lesser Yellowlegs migrate down the middle and more eastern longitudes, rather than along the Pacific coast.

Unlike most of the Greater Yellowlegs that will winter here or in Mexico, Lesser Yellowlegs are on an astonishingly long journey, from where they were hatched in northern Canada and Alaska to the Southern Cone of South America.

Willet *Tringa semipalmata*

LENGTH 15–17 IN.; WING SPAN 26 IN.

Pl. 76

"Will it or won't it?" is not the question, for the Willet's name is an allusion to its obstreperous, panic-stricken vocalizations: "will-will-willet, will-will-willet." One of the most ubiquitous and boisterous of our shorebirds, the Willet is common and conspicuous at most coastal wet spots—sandy beach, mudflat, rocky shore, slough, and marsh—throughout winter.

These are sturdy sandpipers, rather heavy bodied, not as slender and elegant as other tringine sandpipers, with strong, straight bills and gray-blue legs. When a Willet is standing on the shore or foraging, you see plumage that is mostly unmarked gray and a whitish belly. In spring, some dark barring on the wings and body lends some variation to the otherwise dull gray plumage. In flight, the flashing black-and-white wing pattern is unique among shorebirds and readily recognized. Some Willets are present along the coast all year, but the species does not nest here. Instead, it breeds on islands and edges of alkali lakes in the Great Basin as nearby as Mono, Lassen, and Siskiyou Counties.

Migration from nesting grounds to the California coast is nonstop, and we see the first juveniles here in the second week of July.

Whimbrel *Numenius phaeopus*

LENGTH 17–18 IN.; WING SPAN 32 IN.

Pl. 77

One of only two common large shorebirds with distinctly downturned (decurved) bills, the Whimbrel is told readily from the Long-billed Curlew by its black-and-tan head stripes and overall colder, gray-brown plumage. Curlews have plainer crowns and warmer orange-brown feathering.

Fresh-plumaged Whimbrels are graceful birds with a statuesque posture, one of those species that blends in with mixed flocks of large waders (godwits, curlews), barely noticeable to the unaided eye, but magnificent to behold through bright optics. Often seen in small groups, rarely in large flocks, Whimbrels frequent the outer beach as well as tidal flats but may also be found on flooded pastures, salt marsh, or rocky shorelines or flying out over the ocean. Whimbrels nest in northernmost Alaska, migrating to the temperate and tropical coastline for the nonbreeding season. Most winter south of our region, passing through in migration.

Long-billed Curlew ***Numenius americanus***

LENGTH 21–24 IN.; WING SPAN 35 IN.

Pl. 11, 77

Our largest sandpiper, the Long-billed Curlew towers over all of the "peeps," the smallest of the sandpipers. The most distinctive characteristic is the very long, extremely decurved bill, which on some larger adults may reach 9 in. It appears somewhat ungainly, but the shape is identical to the shape of the burrow of the ghost-shrimp, an adaptation that allows for probing deeply in tidal flats and extracting deeply buried prey. (*Numenius*, the genus name, means "crescent moon" in reference to the bird's bill shape.) The upper mandible is slightly longer than the lower, and it ends in a bulbous tip with which the bird is able to grasp prey even as its bill is deeply embedded in the mud. (This ability of the tip of the bill to grasp is called "rhinokinesis" and is a common adaptation of the long-billed sandpipers.) The female curlew is larger than the male, most noticeably in bill length. Except for the bill and the legs and feet (blue-gray for the curlew, blackish for godwit), Long-billed Curlews and Marbled Godwits look very much alike and may roost in mixed-species flocks.

> **The tide rises, the tide falls,**
> **The twilight darkens,**
> **the curlew calls;**
> **Along the sea-sands**
> **damp and brown**
> **The traveler hastens**
> **toward the town. . .**
> **—Henry Wadsworth Longfellow**

Curlews nest in native grassland habitats of the Great Basin and Great Plains and are therefore relatively short-distance migrants, moving to the coast in the nonbreeding season. They seem as at home on dry grassland as on the gooey mud of tidal estuaries. The name *curlew* is mimetic of the bird's plaintive two-note call, mentioned in many poems—perhaps a symbol of mourning, unrequited love, sadness.

SPECIAL STATUS Highly Imperiled on U.S. and Canadian Shorebird Conservation Plans due to loss of native prairie habitat in breeding range in the continent's interior.

Marbled Godwit *Limosa fedoa*

LENGTH 17–19 IN.; WING SPAN 30 IN.

Pl. 77, 78

Like the Long-billed Curlew, the Marbled Godwit is a large, tall shorebird with a warm, orangey-brown plumage. Unlike the curlew, this godwit's bill—4.5 to 6.5 in. long—is gently upturned (recurved) and is bubble-gum pink at the base and black at the tip. The godwit's legs and feet are blackish, while the curlew's legs are blue-gray; this is a useful field mark when a mixed flock of these waders is roosting together with their bills tucked beneath their feathering.

Marbled Godwits are among the common large shorebirds at most coastal wetlands (including open beach) from mid-July through April, and some few nonbreeders stay through summer, sometimes in substantial numbers. This is a gregarious species that tends to forage and roost in rather large flocks, often intermixed with Willets, Long-billed Curlews, and Whimbrels. When foraging on tidal flats, or along the wet sandy beach, Marbled Godwits walk slowly, often in groups, probing deliberately in a relaxed manner, unlike the frenetic smaller shorebirds with whom they may associate. Apparently, godwits tend to return to the same wintering sites each year, a trait termed "high site fidelity" or "philopatry." This tendency may be rather common among migrant shorebirds and migrant birds in general.

Godwits, like curlews, are not spectacularly long-distant migrants, moving from the central prairie provinces, where they nest, to the coast for the nonbreeding season.

Ruddy Turnstone *Arenaria interpres*

LENGTH 9.5 IN.; WING SPAN 21 IN.

There are only two kinds of turnstones. The Ruddy is a holarctic species that occurs along the coasts of all continents except Antarctica. The Black Turnstone is nearctic, nesting only in Alaska and wintering along the immediate coast from Alaska to the tip of Baja and northwest Mexico. Ruddies are fairly uncommon here; Blacks are quite common. Both species have a more plover-like body shape than the other members of the family; in fact, they used to be included in the plover family (Charadriidae) rather than the shorebird family (Scolopacidae) based on superficial similarities, but modern genetics has reassigned the turnstones.

Ruddy Turnstones are spectacular in their harlequin breeding (alternate) plumage, when seen here during migration in spring (April to May) and fall (July to August). This is a shorebird species to look for in late July toward the end of the summer "birding doldrums," as their flashy plumage provides a needed boost of adrenaline. Ruddies are much more cryptically patterned during winter and are relatively scarce, except at Bodega Harbor, where they hang out on piers behind the seafood restaurants and rocks at the Westside Park boat launch. However, Ruddies may be found in some places during high tide, roosting amid flocks of Black Turnstones and Surfbirds.

As their common name implies, they use their closed bill to turn over small rocks, shells, and blades of kelp in search of hiding invertebrates. Ruddies nest in the high Arctic and migrate to southern coastal areas in winter, mostly south of our region.

Black Turnstone *Arenaria melanocephala*

LENGTH 9 IN.; WING SPAN 21 IN.

Pl. 79

Classic rocky shore birds, Black Turnstones forage rocky intertidal habitats at low tide. At high tide, they join communal roosts with Surfbirds, tattlers, and oystercatchers on predator-safe offshore sea stacks. Although the rocky shore is their preferred habitat, Black Turnstones also forage within bays and estuaries, along the cobbled shoreline, or even riffling through marine algae on the upper beach or eelgrass wrack left by the tide at the salt-marsh edge. When foraging, they use their blunt bills to overturn or poke around seaweed or other shore detritus for their favored fare, especially amphipods. Turnstones will also pummel the shells of barnacles, limpets, or mussels to get at the meat. They are a flocking species, usually of a dozen or more birds. The flocks are often quite chatty, their voices a staccato series of throaty, dry rattles.

Black Turnstones nest on or near the coast of western and southern Alaska, moving a short distance southward to the western coast of British Columbia, the United States, and northern Mexico to winter.

Red Knot *Calidris canutus*

LENGTH 10–11 IN.; WING SPAN 23 IN.

The Red Knot is one of those waders named for its fancy breeding (alternate) plumage, but here during winter in nonbreeding dress, knots are *not red*, but drab gray and white.

Knots are uncommon, at best, along the Northern California coast and, except when in bright alternate plumage (in April), blend well with other species (such as dowitchers) and are seldom seen. They are of medium size, without any color or much pattern, and have straight, slender bills. In structure, the knot is similar to the Sanderling, though noticeably larger. The Red Knot in basic plumage could *define* the average, drab shorebird, but notice the narrow white wing stripe and white tail in flight. Preferred habitat includes tidal flats and sandy shoreline, less commonly rocky shore.

Knots are among the longest-distance migrants in the animal kingdom, moving from breeding grounds above the Arctic Circle to wintering areas in southernmost South America. Small numbers winter along the California coast, from Humboldt southward, and breeding-plumaged adults can be seen at coastal sites during spring (April) and fall (August to early September) at almost any shoreline site.

Surfbird *Calidris virgata*

LENGTH 9–10 IN.; WING SPAN 26 IN.

Pl. 80

The Surfbird's distribution is much like that of the Black Turnstone—nesting only in Alaska and wintering along the Pacific coast of North (and South) America. Although allied with the typical shorebirds, the Surfbird is a stouter model, designed more like the plovers and turnstones than the more delicate sandpipers of the tribe Calidridini. This sturdy body shape must serve the Surfbird well in its wave-pounded habitat.

Except for migrants on mudflats in September and April,

Surfbirds stick entirely to rocky-intertidal habitats, often flocking with Black Turnstones, Wandering Tattlers, and Black Oystercatchers. This is one of the less common shorebirds in our region, though perhaps more common in winter in the more northerly counties. The favored habitat is often inaccessible to humans, and the plain gray back blends in with the surf-splashed rocks, making Surfbirds difficult to spot.

Sanderling *Calidris alba*

LENGTH 8 IN.; WING SPAN 16 IN.

Pl. 4, 73, 81

Sanderlings are the pipers most closely related to sand. They are the chunky, pot-bellied little white sandpipers, common on every ocean beach in the region, seen following the receding waves in the swash zone to probe the damp sand, then quickly sprinting to the upper beach for safety from incoming waves. They are foraging for burrowing invertebrates that come to the surface to feed when the wave washes over, mostly abundant mole crabs *(Emerita analoga)*. When roosting on the upper beach, Sanderlings often stand on one leg, a peculiar habit that must serve some physiological function, perhaps energy conservation (thermoregulation).

Studies on the shore of Bodega Bay found that wintering Sanderlings defend small territories in the wave-washed zone along the shoreline of the sandy beach (Myers et al. 1979).

Like most *Calidris* sandpipers, this is a long-distance migrant that is highly faithful to its wintering beaches. Numbers are highest in winter (August through April), but small numbers remain on our beaches all summer long. Sanderlings have one of the broadest wintering ranges of the sandpipers, stretching from southern British Columbia southward to southern Chile, spanning 100 degrees of latitude.

There were more Sanderlings along our beaches just 20 years ago, but like Snowy Plovers, Sanderlings have trouble competing for space with humans and dogs on the ocean beach. "Regional populations are in rapid decline, with the apparent cause being habitat degradation and increasing recreational use of sandy beaches" (Birds of North America Online).

Dunlin *Calidris alpina*

LENGTH 7–8 IN.; WING SPAN 13–15.5 IN.

Pl. 83

Following nesting season in the Arctic, most species of southbound migrant sandpipers appear on the coast in the last week of June or in July. The first waves of each kind are adults, followed by the first juveniles three or four weeks later. The Dunlin is a bird of a different feather, remaining on the tundra while molting into basic plumage. The first "normal" migrant Dunlin does not reach California until mid-September, 10 weeks after most other species. Once the masses arrive in October, the Dunlin becomes one of the more abundant shorebird species at coastal lagoons and estuaries.

Although possibly mistaken for a smaller "peep," the Dunlin is clearly bigger than Least and Western Sandpipers, with uniform gray-brown upper parts in winter dress and obviously decurved, or droopy, bills. Dunlins show a striking difference between basic winter plumage and alternate breeding plumage. Winter dress is a drab ("dun") gray above, suffusing to white below. Juveniles, "birds of the year," show some brown edging to the back feathers. By spring Dunlins may be seen in their dramatic breeding plumage, with black underparts and rich shades of brown patterning on the back and wing coverts, which have earned them the alternate name Red-backed Sandpiper.

In a mixed flock of small shorebirds, Dunlins often forage belly deep in water, submerging their heads to probe, whereas Western Sandpipers are at the water's edge, and Least Sandpipers tend to forage on higher, drier ground. The careful observer will soon notice the distinct bill shapes of the three species, adaptations to their differing foraging strategies. Dunlins may appear more hunchbacked when foraging than the other small shorebirds.

Wherever Dunlin flocks gather, a Merlin is likely to be found, posed on some distant perch, watching the flock, waiting for an opportunity to strike.

Baird's Sandpiper *Calidris bairdii*

LENGTH 7–8 IN.; WING SPAN 17 IN.

Baird's Sandpipers, like Pectoral Sandpipers and Lesser Yellowlegs, are very uncommon along the coast and have a strange seasonal and distributional story. With very few exceptions, Baird's Sandpipers occur on the California coast *only* in fall (July to October), and all of them are "birds of the year," hatched the previous spring on Arctic breeding grounds. It is very rare to see an adult here, even in fall, or a Baird's Sandpiper of any age in spring. Adults have a narrow migratory route down the center of the continent, whereas young birds, for reasons unknown, move on a much broader front.

Spencer Fullerton Baird was one of the five most important contributors to early American ornithology. Baird's Sandpiper was the last bird described and artistically documented by John James Audubon and was the first of several species named for Baird.

Least Sandpiper *Calidris minutilla*

LENGTH 5–6 IN.; WING SPAN 10 IN.

Pl. 82

Leasts, as the name implies, are tiny sandpipers—in fact, the smallest worldwide. Their yellow legs and feet are unique to nearctic "peeps," but their tiny size and mouse-brown dorsal coloration (white underneath) will help identify them at a distance. Although the yellow legs are a reliable field mark, when Leasts have been wading in mud, the legs can appear black. The Least's bill is finer and shorter than that of the Western Sandpiper with which it often associates.

Although Leasts sometimes forage on open mudflats with Dunlin and Western Sandpipers, they usually prefer higher and drier situations, often foraging above the tide's edge and roosting way back in the pickleweed or detached eelgrass wrack during high tides. Leasts seldom or rarely participate in "flock swirling" as do other shorebirds when there is a falcon attack. They just crouch low, blending with the substrate, their cryptic dorsal coloration making them all but invisible from above. Although quite common, Least Sandpipers are not as abundant as the other two most common small shorebirds, Dunlin and Western Sandpipers, although at some sites with muddier substrates, they may predominate.

Western Sandpiper *Calidris mauri*

LENGTH 5.5–6.5 IN.; WING SPAN 12–14 IN.

Pl. 82

Along with Dunlins and Least Sandpipers, Westerns are in the generic category "peeps," the smallest of the sandpipers. Of the peeps, Westerns are the whitest (but see Sanderling) when in basic plumage, November through March.

Like the other peeps, Western Sandpipers are tidal-flat specialists, rapidly probing the exposed mud when the tide is low, roosting in dense flocks when the tide is high. They also forage on the outer sandy beach when the flats are unavailable.

At most places and times along the coast, the Western is also the most common peep, especially in tidal estuaries. San Francisco Bay is a major wintering ground for Westerns and has been designated as an estuary of International Importance as part of the Western Hemisphere Shorebird Preserve Network, for its value to shorebirds in general and to this species in particular. During migration, the numbers of Westerns swell even more, with tens of thousands gathering at favored sites. Other major refueling stations include Elkhorn Slough and outer coast estuaries (Bolinas Lagoon, the Point Reyes esteros, Bodega Harbor, and the Humboldt Bay complex).

Those animated clouds of shorebirds swirling over mudflats in response to a hunting Merlin or Peregrine Falcon are mostly Western Sandpipers. After September, Dunlin may join those flocks, but the smaller peeps—Least Sandpipers—prefer to crouch and freeze rather than panic and fly.

Pectoral Sandpiper *Calidris melanotos*

LENGTH 8–9 IN.; WING SPAN 13.5–17.5 IN.

Pectorals are like Baird's Sandpipers in their seasonal and age-specific distribution on the California coast. They pass virtually *only* in fall, and the few that do pass are all juveniles hatched the previous spring. They are on their way to winter in the Southern Cone of South America and will return to the Arctic by a different route the following spring. "Pecs" are most often seen at the edges of still ponds, at lagoons and sloughs, in salt-marsh habitats, or even in damp pasture, but rarely out on the tidal flats.

The common name alludes to the area of the chest where the promiscuous male inflates air sacs in an elaborate display on the northern breeding grounds. In the same pectoral area, the streaked neck and chest abruptly demarcated from the white belly provide an excellent field mark. Pectoral Sandpipers have good posture; they hold themselves tall and statuesque. The nearly identical Sharp-tailed Sandpiper *(Calidris acuminata)*, a Siberian to Australian migrant, is extremely rare in our region.

Long-billed Dowitcher ***Limnodromus scolopaceus***

LENGTH 10 IN.

Short-billed Dowitcher ***Limnodromus griseus***

LENGTH 9.5 IN.

Pl. 10, 84

The two species of dowitchers are very difficult to separate in the field, so they are treated together here. In fact, they were not recognized as separate species until the mid-20th century.

The dowitchers are most often seen in small, tightly packed flocks, probing actively ("stitching") with a "sewing-machine motion," belly deep in shallow water or at the edge of the tide. Despite their respective names, bill length is not a very reliable field mark when trying to distinguish between these two species. However, behavior may be helpful. The sewing-machine-like foraging behavior is similar in both species, but when "stitching," Short-billeds tend to appear flat backed, while Long-billeds seem more humpbacked. The calls of the dowitchers differ, and voice recognition may be the easiest way to identify the birds. Long-billeds speak with a harsh "keek" or "keek-keek-kek." Short-billeds whistle a gentle "too" or "tu-tu-too." Long-billeds are chattier than Short-billeds, so a dowitcher flock that is silent when feeding is probably a group of Short-billeds. Once fall migration is over and all are settled for winter, Short-billeds are found only on the largest tidal estuaries (San Francisco Bay) while Long-billeds might be at any coastal wetland. Long-billed Dowitchers are the more common of the two species in winter, though both occur here. All dowitchers are scarce north of Marin County in winter.

Plumage differences, though subtle, are most pronounced in juvenal plumage (from June into October) and for a brief period during spring migration when birds have attained alternate breeding plumage. In fall, juvenile Short-billeds are more highly patterned than are Long-billeds, especially on the wing coverts and tertials that have orange margins and interior barring. During spring migration (April), sizable flocks of Short-billed Dowitchers move through the coastal estuaries, already showing the warm russet shades of breeding plumage that may help distinguish them.

Habitat preferences may differ subtly between the two species, but they also gather in mixed flocks, if only to challenge and confound the curious naturalist.

Wilson's Snipe *Gallinago delicata*

LENGTH 10 IN.

Pl. 85

This snipe, formerly called Common Snipe, is reclusive and cryptic, usually concealing itself in the grasses and sedges of freshwater marshes, ponds, or slow-moving streams, and occasionally in salt-marsh habitats. Wilson's Snipe is easy to overlook; it is most often encountered when it gets flushed from a few feet away and bursts into the air calling an emphatic, nasal "skaaaip," flying off rapidly in an erratic zigzag pattern. In flight the silhouette is dowitcher-like, but no other shorebird has the erratic flight path of the snipe.

Wilson's Snipes winter along the coast but do not nest here. The breeding grounds are in the interior, in wet meadows and marshes east of the Sierra-Cascade cordillera. The "winnowing flight" performed above the nesting grounds is a dramatic sky dance, a spectacle that rivals the breeding displays of the hummingbirds, and is similar in the way that it produces a loud sound with the vibration of the tail feathers.

The word *snipe* is a variant of *snout*, in reference to the extraordinary long bill of this bird, used for probing deeply in soft substrate. The tip of the bill is flexible and densely enervated, affording the bird the ability to grasp prey at depth. Apparently earthworms are a favored food, but larval insects, mollusks, and crustaceans are also taken.

Phalaropes

Phalaropes are unique members of the sandpiper tribe. *Phalaropus* translates to "coot foot" in Greek, and all three species have evolved lobed toes for swimming and foraging in aquatic habitats. Like the Spotted Sandpiper, phalaropes exhibit a gender-role reversal (reverse sexual dimorphism). During spring the female is more brightly patterned than the male; the female establishes and defends a territory and then selects a male for mating. After egg laying, the female drifts away from the territory, and all domestic responsibilities (incubation and chick fledging) belong to the male.

While swimming in shallow water, phalaropes often spin in tight circles, apparently to draw brine shrimp and other small invertebrates to the water's surface, for easy forage. There are three phalarope species in the world, and all occur on the Northern California coast during migration, spring and fall.

Red Phalarope — *Phalaropus fulicarius*

LENGTH 8 IN.

Highly pelagic during both spring and fall passage, Red Phalaropes are rarely seen on the shoreline of the coast, but when blown inshore by gale-force northwest winds, numbers can be impressive. Tens of thousands migrate 10 to 50 miles offshore and are easily seen from ocean-going boats, but this species never willingly goes to land except to breed, in the high Arctic.

Most Red Phalaropes winter south of the equator, but a few remain here most years and are sometimes seen from shore. This is the only one of the three species to be expected along the coast in winter.

Red-necked Phalarope — *Phalaropus lobatus*

LENGTH 7 IN.

Pl. 86

After nesting in the low Arctic and subarctic, Red-neckeds migrate southward over the sea, miles offshore, as the Red Phalaropes do. Unlike Reds, many Red-neckeds also pass onshore and up and down the continent's interior. (During September they stop at Mono Lake for refueling, eating pupating alkali flies. Single-day surveys there in the mid-1980s averaged 50,000 birds.)

This is the most common phalarope found onshore along the coast. In fall especially, Red-necked Phalaropes predictably appear on calm bodies of water, usually in small numbers, actively foraging on the water's surface, occasionally along the shoreline. They have a relatively delicate stature, a thin bill, and a long neck and are especially fleet and agile on the wing.

Wilson's Phalarope ***Phalaropus tricolor***

LENGTH 8.5 IN.

Wilson's Phalarope is endemic to the Americas, nesting on plains and Great Basin habitats in the north and wintering at ponds on the pampas of Argentina and saline lakes in the Andes. This is the rarest of the three phalarope species along this coast. Most of these birds migrate over the continental interior (over 100,000 may be present at Mono Lake, late June and July), but individuals or small groups are detected annually along the coast in July and August, more rarely in April and May.

Skuas and Jaegers

Family Stercorariidae

Skuas and jaegers nest in the high latitudes but are highly pelagic in the nonbreeding season, occurring well offshore along the Northern California coast. These are athletic pirates and predators, with deep chests, long and pointed wings, and a distinctive flight as they pursue other seabirds—especially gulls and terns—from whom they steal prey. The outer flight feathers (primaries) are white at the base, in varying degrees among species; the central tail feathers of the jaegers are elongated and uniquely shaped in each of the three species. The beak is rather short and stout, with a raptor-like hook at the tip. Unlike the gulls and terns, this family has females that are larger than males. Like the gulls, these are long-lived birds: breeding age is not reached for three to seven years in jaegers,

A Parasitic Jaeger chasing an Elegant Tern is a fairly common high-energy spectacle along the coast in the fall. *K.H.*

even later in the skuas. Only the Parasitic Jaeger *(Stercorarius parasiticus)* is seen regularly from shore; the Long-tailed Jaeger *(S. longicaudus)*, the Pomerine Jaeger *(S. pomarinus)*, and the South Polar Skua *(S. maccormicki)* are truly pelagic seabirds, migrating along this coastline far offshore.

Parasitic Jaeger ***Stercorarius parasiticus***

LENGTH 16 IN.; WING SPAN 46 IN.

Jaegers are pirates that chase other kinds of fish-eating birds, harassing them into giving up their meals. *Jaeger* means "hunter" in German, and while that may be true during the nesting season when they eat eggs, longspurs, and voles, *pirate* is the right word for their behavior during the nine months at sea. After nesting on the Arctic tundra, Parasitic Jaegers range widely along the Pacific coast all the way to South America, keying on migrating terns as they pass our shores.

Of the three jaeger species, the Parasitic is the one that occurs closest to shore, where they chase after Forster's and Elegant Terns. From late August into October, watch for the angular, dark jaeger slicing along and chasing a potential victim.

Alcids

Family Alcidae

Auks, also known as alcids, are seabirds restricted to the Northern Hemisphere and are the ecological equivalent of the Southern Hemisphere's penguins. Like penguins, they are mostly black above and white beneath, with compact bodies, and webbed feet set far to the stern. Unlike penguins, alcids can fly through the air, as well as underwater. These are plunge divers, pursuing schooling fish, krill, and zooplankton in the cool waters of the California Current, often far offshore. The bill shapes of the several species have evolved as they have exploited specific types of food, from the large remarkable hatchet-shaped bill of the Tufted Puffin to the proportionally tiny blunted beak of the Cassin's Auklet. Most alcids go to land only to nest, colonially, in burrows, in rock crevices, or on cliff ledges on islands or inaccessible headlands. One species, the Marbled Murrelet, nests on the mainland, in old-growth conifer forests. The Northern California coast, with its productive upwelling ocean waters, hosts 10 species, 6 of which are common enough in our region to be treated here.

Alcids have suffered dramatic population declines historically, victims of egg thievery by humans, depredations by introduced rodents and burgeoning gull populations, oil spills, and the vagaries of sea-surface temperatures associated with El Niño–La Niña cycles. Fortunately, most species in our region have proved resilient and rebounded with the protection and management of nesting sites, most notably the Farallon National Wildlife Refuge, 27 miles west of the Golden Gate, managed jointly by the U.S. Fish and Wildlife Service and Point Reyes Bird Observatory Conservation Science (renamed Point Blue Conservation Science in 2013). Under this careful management, South Farallon Island supports the largest seabird breeding colony south of Alaska and hosts 30 percent of California's nesting seabirds. Protection of the island was strengthened by the establishment first of the Gulf of the Farallones National Marine Sanctuary (1972) and then the Cordell Bank National Marine Sanctuary (1989).

The first two species treated below are easily seen from land along the coast. The two murrelets also occur in nearshore waters, but they are rather rare and any sighting of either is noteworthy. The auklets and the puffin are truly pelagic but may be seen during summer at their few coastal nesting sites, with luck and persistence.

Common Murre ***Uria aalge***

LENGTH 6–17 IN.; WING SPAN 25–28 IN.

Pl. 12, 87

One of the most common seabirds in the Northern Hemisphere, the Common Murre thrives in the cold waters of the California Current. Murres are large members of the alcid family and very deep divers, pursuing fish and marine invertebrates at depths of 300 ft or more. Highly colonial, they nest side by side on steep cliffs on islands, sea stacks, and promontories inaccessible to mammalian predators. Like many other seabirds, they build no nests, and for each pair a single large

Common Murre chicks fledge when quite small and follow the adult, usually the father, through the ocean swells, often near shore. *K.H.*

egg is laid each year on a ledge or rock crevice. The parents share incubation duties, which last about one month, and then feed the chick regularly for the next three to four weeks until fledging.

As a tiny chick, a Common Murre jumps from its birth ledge into the sea, where a parent, usually the father, quickly joins it. The two swim away and for the next 12 weeks the adult feeds and cares for the youngster until it can fend for itself. During this period, the fully feathered but miniature murre might be mistaken for a murrelet, an alcid that is smaller than a full-grown murre. The duration of parental attendance to its chick at sea is uncertain but is believed to continue for one or two months until the chick is self-sufficient.

Murres are often seen from shore, singly or in small flocks. In fall, a parent being followed by a hungry chick is a common sight. The largest nesting colony in California is at the Farallon Islands, where nearly a quarter of a million nesting birds were estimated in the years 2005 to 2010. Other major colonies are at Point Reyes (Marin Co.), Green Rock, Flatiron

Rock (Humboldt Co.), and False Klamath Rock and Castle Rock (Del Norte Co.).

Though still common, the murre population has faced devastating obstacles: Through the late 19th century, their eggs were harvested to feed the gold-crazed humans, which extirpated some colonies. Gillnetting in the 1970s and early 1980s killed hundreds of thousands. El Niños can raise sea surface temperatures, driving prey species to deeper levels and starving young and adult murres and causing total reproductive failure. Oil spills are tougher on murres than on any other bird. On the water, flocks of murres drift on currents or with the wind, the same factors that influence oil slicks, so the birds end up coated in viscous hydrocarbons, a lethal fate.

Pigeon Guillemot ***Cepphus columba***

LENGTH 12–14 IN.; WING SPAN 23 IN.

Pl. 88

Like the Common Murre, the Pigeon Guillemot is endemic to the North Pacific and has the classic body shape of an alcid. Guillemots are summer birds here, nesting in burrows and crevices on steep headlands and offshore rocks. They are creative in the selection of nest sites, often raising young under wharves and coastal buildings, even on abandoned boats at anchor. The Farallon Islands, off San Francisco, support one of the largest nesting colonies, but they can be seen at most headlands or coastal wharfs from Monterey to Del Norte County.

The tomato-red legs and feet of the adult are color-coordinated with its mouth lining, which it displays readily in courtship on the nesting rocks. It is quite amazing to see these colors during the birds' squealing and trilling songs. A pair will often roost atop a prominent outcropping of a sea stack, uttering high-pitched trills, or may be seen circling in front of a headland or cliff face above the breakers, or diving in rough seas.

After nesting on the California coast, most Pigeon Guillemots migrate north to spend winter with their congeners from Puget Sound to the Bering Sea. The few that winter here are "young of the year." Since adults arrive and depart in black-and-white (alternate) plumage, we never see them in the much whiter basic plumage.

Marbled Murrelet ***Brachyramphus marmoratus***

LENGTH 9 IN.; WING SPAN 16 IN.

Pl. 89

Called "fog lark" by loggers, this little seabird has the most unusual lifestyle of any of its family members. It forages exclusively in rather shallow marine waters within a mile or less of shore, but commutes inland to nest in mature, old-growth conifer forests from British Columbia southward to Santa Cruz County. This murrelet does not build a nest, but lays a single egg amid mossy clusters on a large horizontal "platform" branch. During the nesting season, Marbled Murrelets leave the coastal waters to attend their nests only in the twilight hours, flying very swiftly, usually at canopy level or below the fog bank, inland to nesting trees. Incubation is shared by both sexes, with duties at the nest exchanged every 24 hours at dawn. While one bird incubates, the other forages along coastal waters. Chicks grow rather quickly and leave the nest at about four to five weeks of age, apparently flying directly to the ocean (Nelson 1997).

The sexes look alike but have unique breeding and winter plumages. Their summer (breeding) plumage is a cryptic sooty brown with rusty highlights; they are more easily spotted on the water in their black-and-white winter plumage. When seen from shore, Marbled Murrelets are often in pairs or small groups, floating and foraging for small schooling fish just beyond the breakers along rocky shorelines or sandy beaches or in protected coves—more commonly in the more northerly counties. Logging of coastal forests has destroyed and fragmented large tracts of the nesting habitat of the Marbled Murrelet and poses an ongoing threat to the population. Predation of nest sites, especially by corvids, has also been identified as a cause of the species' decline. The highest risk of predation was documented in areas close to humans (within 0.5 mi), including along suburban edges and at campgrounds, dumps, and other areas of development, where human food sources attract predators, especially corvids. Steller's Jays may pose the greatest threat to murrelet reproductive success, especially in the southern portion of their range.

Although Marbled Murrelets are never easy to find, some

fairly reliable places to see them from shore are Point St. George, the mouth of Humboldt Bay (North Jetty), Patrick's Point State Park, and Waddell Creek (Santa Cruz Co.).
SPECIAL STATUS Federally Threatened; state Endangered.

Ancient Murrelet *Synthliboramphus antiquus*

LENGTH 10 IN.; WING SPAN 17 IN.

This small, sturdy seabird disperses from its northerly nesting areas—southern Alaska to British Columbia—into our region in winter. The stout, pale bill is designed for capturing its primary prey, euphausid shrimps. The Ancient Murrelet tends to stay farther offshore than the Marbled Murrelet, though both species may occur together and may be seen from coastal promontories in fall and winter. In addition to its compact "GISS," the most reliable field marks are the gray back contrasting with a black crown and striking white patches on the sides of the neck. With a rare excellent look, you may notice the pale bill. In flight, the Ancient Murrelet has a gray rump, whereas the Marbled Murrelet shows obvious white on the sides of the rump in nonbreeding plumage.

Cassin's Auklet *Ptychoramphus aleuticus*

LENGTH 9 IN.; WING SPAN 15 IN.

Pl. 90

Although considered one of the three most abundant seabirds in Northern California, and distributed broadly from Alaska to Baja, Cassin's Auklet is a bird rarely seen from shore. This uniformly mouse-gray alcid is chunky, slightly larger than the Ancient Murrelet, and noticeably smaller than the Rhinoceros Auklet, which is similarly gray colored. Its flight is strong and fast, but this body is designed primarily for diving and the pursuit of euphausid shrimps, larval squid, small fish, and crabs. For nesting, the Cassin's will dig a burrow in the ground with its strong, clawed feet but will also use natural cavities—rock crevices and caves—opportunistically, and even artificial nest boxes. At night, the remote nesting colonies echo with the cacophonous calling of nesting auklets, all sounding off in unison, a sound that has been described as akin to a frog pond chorus, a cricket symphony, or the squealing of piglets.

Cassin's Auklet pairs lay a single egg, then both parents share incubation duties for a relatively long period, 37 to 57 days. This lengthy incubation period allows the chick to emerge from the egg well developed and covered with a full coat of fluffy down feathering. The chick will stay in the birth burrow for about one month and by the end of that period will have attained juvenal plumage, no longer downy. The chick will leave the nest for the ocean at about six weeks of age, and forage independent of adults.

Cassin's Auklet is highly sensitive to ocean water temperatures and responds to high productivity associated with cold-water periods by nesting twice in a season, a unique ability within the auk family. The largest colonies in our region are on South Farallon Island off San Francisco and on Castle Rock, Del Norte County. The Farallon Islands support the state's largest nesting colony, but numbers apparently fluctuate widely, with estimates ranging from 10,000 to 30,000 individuals over the last several decades. Numbers have declined in recent decades in response to changing ocean conditions, and perhaps in response to an increase in baleen whale populations, especially Blue Whale *(Balaenoptera musculus)* and Humpbacked Whale *(Megaptera novaengliae)*, direct competitors for the same prey (Ainley and Hyrenbach 2010). Perhaps the greatest threat is depredation of eggs, chicks, and adults by mammals (cats, rats, and mice), which have been introduced to the nesting islands, and by Western Gulls, whose numbers have been subsidized by human refuse.

SPECIAL STATUS Bird Species of Special Concern, priority 3.

Rhinoceros Auklet *Cerorhinca monocerata*

LENGTH 15 IN.; WING SPAN 22 IN.

A puffin-like alcid of northern waters, the Rhino has a few nesting outposts in California—the Farallon Islands (San Francisco Co.), Castle Rock (Del Norte Co.), and possibly Trinidad Head, Sugar Loaf, and Green Rock. It is oddly scarce in Oregon, where heavy predation by the Great Horned Owl is suspected on the few nesting sites (Marshall et al. 2003). Like the Cassin's Auklet and Tufted Puffin, the Rhino is rarely seen from land; though unlikely, a sighting is possible. Thousands winter in Monterey Bay and other areas where deep waters approach the coast, but they occur "mainly in continental-shelf waters and at the shelf break, overlapping broadly with distribution of Cassin's Auklet" (Gaston and Dechesne 1996).

Tufted Puffin *Fratercula cirrhata*

LENGTH 15–17 IN.; WING SPAN 25 IN.

Pl. 91

Wearing a clownish mask and carrying a huge multicolored bill, puffins are rarely seen from shore except off Mendocino, Humboldt, and Del Norte Counties. There, with use of a good telescope, puffins may be spotted at their burrow entrances on rocks at Castle Island (Del Norte Co.); Puffin, Green, and Flatiron Rocks (Trinidad, Humboldt Co.); and Goat Rock (Mendocino Co.). Few sights are as enchanting or as rarely seen as an adult puffin landing at its burrow entrance with a half dozen anchovies or small squid dangling from its huge multicolored bill.

South Farallon Islands, the most important nesting site south of Cape Mendocino, held several thousand nesting birds in the early 1900s, but that colony has declined to only about 100 pairs in recent years, mirroring a trend throughout the region. Relatively warm-water periods over the last several decades have precipitated a dramatic decline in puffin populations, not only here, at the southern edge of their distributional range, but also off the coasts of Oregon, Washington, and British Columbia. (Predation pressure at nesting colonies by recovering populations of Bald Eagles is a contributing factor to the decline in Oregon.)

The Tufted Puffin is the "most pelagic of the alcids." During the nesting season, it forages far offshore, mainly in waters over the continental shelf and the continental slope. Indeed, few puffins winter over these deep waters of the Pacific at California's latitude; most probably go north, like guillemots, to colder, more reliably nutritious waters.

SPECIAL STATUS Bird Species of Special Concern, priority 1.

Gulls and Terns

Family Laridae

As a family, the larids are a conspicuous coastal presence, known to most of the public simply as "seagulls." All members of the group are superficially similar: All have webbed feet, and males are generally larger than females. Both sexes attend the nest. Most have white bodies in adulthood, nest colonially, and are gregarious the rest of the year. Close observation reveals a family of extraordinary diversity. Some gulls (subfamily Larinae) resemble terns (subfamily Sternidae) structurally, and some terns are as large as some gulls. The larger gulls have eclectic diets (opportunistic omnivores), are rather long-lived (more than 20 years), and take three to five years to mature into adult plumage. The terns are primarily fish eaters and forage by plunging headlong into the water. Some of the smaller gull species (e.g., Bonaparte's and Sabine's Gulls) are quite tern-like in structure and behavior. Ten species of gulls and five tern species are found annually in our region, ranging in size from the diminutive and graceful Least Tern to the large and powerful Glaucous-winged and Western Gulls. Among these larger "white-headed" gulls, DNA analysis reveals a fairly recent evolutionary divergence among some species, so hybridization is common and species relationships are "unsettled."

Identification of most of the terns commonly encountered in our region is relatively straightforward and well illustrated by most field guides. The gulls pose their own set of challenges, and their identification is often avoided by novice birders. The younger gulls tend to wear darker plumages as they age toward adulthood, making identification even more difficult. The adult plumage is a good place to start with most species, because each species is structurally distinct or has unique combinations of leg and eye color, bill color, shape and pattern, and wing pattern. Once the adults are recognized, it becomes somewhat easier to sort out the gallimaufry plumages of immature birds.

Black-legged Kittiwake *Rissa tridactyla*

LENGTH 17 IN.; WING SPAN 36 IN.

The lovely, graceful kittiwakes are pelagic gulls, common offshore in winter but seldom seen from land. Some years, however, large incursions deploy to mainland harbors, jetties, and river mouths, delighting birders with their graceful elegance. Occasionally during a winter storm surge, individuals are blown or wander nearshore and can be seen from the beach.

Most similar in appearance to the much more common Bonaparte's Gull (especially true of the juveniles of both species), the kittiwake can be separated by the wing pattern and the cleft tail.

Bonaparte's Gull *Chroicocephalus philadelphia*

LENGTH 13 IN.; WING SPAN 34 IN.

A small and dainty gull, Bonaparte's is gregarious—usually arriving in homogeneous flocks of its own kind or mixed with terns, usually Forster's. Most of its migration is offshore, but the species may be seen at special places along the coast. "Boneys" are rare in winter north of Marin County.

Bonaparte's Gulls are so small and their flight is so buoyant that they are more likely to be mistaken for terns than other kinds of gulls. They don't plunge dive, as do terns, but pick their food from the water's surface while flying. Unique among the gulls, the Bonaparte's nests in trees in Alaska and western and central Canada.

Heermann's Gull *Larus heermanni*

LENGTH 19 IN.; WING SPAN 50–52 IN.

Pl. 92

Perhaps the easiest gull to identify, at least in adult breeding plumage, Heermann's Gull is a handsome bird with a velvety gray back, red bill, and white head, though the head becomes mottled in basic plumage. Playing Sancho Panza to the Brown Pelican's Don Quixote, these tagalongs collect most of their food by picking up fish scraps and leftovers following plunge dives by the large pelicans, or even stealing the fish directly from pelicans' pouches. However, they are not always pirates. Occasionally Heermann's Gulls forage along the wave-washed beach, picking up small crabs or other items from the veil of water, much as Sanderlings or Willets do. The Heermann's Gull and the Brown Pelican plus the Elegant Tern ("the three amigos") come north after breeding in Mexico to forage the cold, nutrient-rich waters of the California Current. Heermann's Gulls are usually evident all along the coast from July through October and are scarce or absent the rest of the year, depending upon water temperature, the presence of small schooling fish, and thus, Brown Pelicans. However, Heermann's Gulls are *always* present from Point Pinos in Monterey County south.

Most of the world population of Heermann's Gulls nest on Isla Raza in the Gulf of California, a sanctuary protected by the Mexican government, but several extralimital attempts at nesting have been made in Northern California. Nests at Alcatraz Island in San Francisco Bay and Año Nuevo Island were depredated by Western Gulls. A few nests have succeeded and fledged young at Robert's Lake, Monterey County.

Mew Gull ***Larus canus***

LENGTH 15–16 IN.; WING SPAN 43 IN.

Pl. 93

To identify most gull species, you must gather several field marks (leg color, bill size, back shade, etc.) before naming the birds. With some experience, only one look at the profile of the head is required to identify the Mew Gull. The name *Mew* refers to the gentle voice of this gull, but the old name, Slender-billed Gull, is more helpful for visual identification. The bill is delicate and small (unmarked in adults), the head is rounded and large, and a dark eye lends the bird a gentle or "cute" look compared with the bulbous-billed, slope-headed, and often wild-eyed look of most other gull species. (Facial expression, or aspect, is an underused identification aid in most field guides.)

Mew Gulls usually begin to arrive from the north in the third week of September and are abundant all along the coast by the end of October. Following soaking rains, large numbers gather in fields to harvest flooded earthworms and other invertebrates. Mew Gulls are rare inland from the coast.

Ring-billed Gull *Larus delawarensis*

LENGTH 17–18 IN.; WING SPAN 48 IN.

Pl. 94

At some point while growing up, most of the local gull species develop a black "ring" around their paler bill, so that alone is not a decisive field mark of the Ring-billed Gull. Better are its small size, pale-gray back, yellow legs and feet, and glaring yellow eyes (irises).

Ring-bills are common all along the California coast in winter but rarely within the large, mixed flocks of gulls assembled on beaches. Ring-billed flocks will gather on open fields, especially baseball diamonds and soccer fields, during inclement weather. They tend to stay away from the larger species—at least more than snapping distance. Ring-bills are shameless scavengers, the most common species begging and picking up crumbs at picnic areas and parks where humans gather. Ring-bills prefer sandy beaches and freshwater lakes; they are rather rare along rocky-shore habitats and do not wander far offshore.

California Gull *Larus californicus*

LENGTH 21–22 IN.; WING SPAN 54 IN.

Pl. 95

Most California Gulls nest in colonies on island lakes in the Great Basin. A large population at Mono Lake was a focal point in one of the greatest conservation triumphs of all time. Because the Los Angeles Department of Water and Power diverted streams that flowed into Mono Lake, the lake level dropped. This caused the gull nesting islands to be attached to the mainland shore, allowing mammalian predators (coyotes) to devastate the colony. Many people felt that was unacceptable and after a long legal battle were able to restore the lake levels and save the gulls.

The California Gull displayed a remarkable ability to rebound. In the late 1970s, a nesting colony was established in southern San Francisco Bay. Over the ensuing decades, the San Francisco Bay Bird Observatory documented the remarkable growth of that colony, from 50 breeding birds in 1980 to over 43,000 gulls in 2009. This dramatic increase in gull numbers in the bay has caused concerns about impacts on other locally nesting waterbirds—especially the Western Snowy Plover, a threatened species vulnerable to egg and chick depredation by an aggressive opportunist.

At many places this is *the* most common gull along the coast in winter; it is also numerous offshore at feeding frenzies and following fishing boats for scraps. The adult is distinctive with its yellow-green legs (sometimes appearing light blue) and its black-and-red bicolored bill. Before transitioning into adult plumage, the first- and second-cycle California Gull might be mistaken for the Ring-billed Gull, but note the larger size, the heavier bill, and, in profile, a flatter forehead than either the Ring-billed or the Mew Gull has. If plumage gets too confusing, use general shape and structure to identify the California Gull. Consider it the midsized gull along the coast and identify it by default (but see Thayer's Gull). The Western, Glaucous-winged, and Herring Gulls are all larger, bulkier birds with thicker bills and broader wings.

Herring Gull ***Larus argentatus***

LENGTH 24–27 IN.; WING SPAN 58 IN.

Pl. 97

Herring Gulls are only present from the end of September into early April. Compared with the California Gull and the Ring-billed and Western Gulls, Herrings are uncommon, although large numbers may appear at herring spawns in the larger bays. They are the only large (pink-legged) gulls to be found in Northern California's interior. If you see a gull other than a Ring-billed or California Gull in the Central Valley in winter, it is likely a Herring. On the coast, Herrings often roost in mixed-species flocks, especially with the other large pink-legged species. The streaked head and nape of the adult is distinctive, but that characteristic may also appear in the rather common hybrids of Western and Glaucous-winged Gulls, sometimes called "Olympic Gulls."

Thayer's Gull *Larus thayeri*

LENGTH 21–23 IN.; WING SPAN 55 IN.

Pl. 96

Thayer's Gull is a virtual enigma, and probably the most frequently misidentified bird in North America. Rather, other birds—hybrids of Western and Glaucous-winged Gulls mostly—are often mistaken for Thayer's. Even seasoned observers grapple with the gull identification, and Thayer's Gull offers a humbling challenge. More than identification problems, the question of the species status of Thayer's Gull has long confounded taxonomists: is it a separate species, a subspecies of the Iceland or Herring Gull, or something in between that defies category? Despite a history of taxonomic uncertainty, Thayer's Gull is currently accepted as a distinct species on the American Ornithologists' Union's Check-list of North American Birds.

For a large larid, the Thayer's is a relatively delicate gull (the Mew Gull of the "pink-legged" gang), and those that visit the Northern California coast travel here from east-central Arctic Canada, perhaps nonstop.

Western Gull *Larus occidentalis*

LENGTH 24–27 IN.; WING SPAN 48–58 IN.

Pl. 97

Strictly coastal, the Western Gull is almost never found inland. Westerns, along with Heermann's Gulls visiting from Mexico, are the most frequent gulls present during summer. Along the Northern California coast, active nests can be seen on coastal headlands, sea stacks, and even piers and abandoned barges in the San Francisco Bay. Approximately 30 percent of the world population nests on Southeast Farallon Island, with 15,000 to 22,000 breeding adults estimated from 1970 to 2010. Declines in reproductive success in recent years are thought to be related to variations in the water temperatures of the California Current.

Western Gulls are separated into two subspecies: the larger and paler *L. o. occidentalis* that nests from the Monterey Peninsula north, and the smaller and darker-backed *L. o. wymani* to the south. The northern group interbreeds extensively with the Glaucous-winged Gull at the northern edge of its range, centered near the mouth of the Columbia River, and hybrids occur regularly in our area, making the challenges of gull identification even more daunting (pl. 97). As in the other large white-headed gulls, males are noticeably larger than females.

Glaucous-winged Gull ***Larus glaucescens***

LENGTH 24–27 IN.; WING SPAN 47–56 IN.

Pl. 97

Glaucous-wingeds are a paler version of the Western Gull and, in fact, may be merely a pallid northern subspecies of the Western Gull. "Pure" Glaucous-wingeds become very common here from late October through March, and are rare in summer. Westerns are relatively more common in the southern regions of California; Glaucous-winged Gulls become more common to the north. Intergrades between the Glaucous-winged and Western are also present during that time period and are sometimes misidentified as Herring or Thayer's Gulls.

A few Glaucous-winged Gulls that lack the energy or fitness to migrate also lack energy to molt, their feathers wearing to nearly white. Each summer, these very light individuals are mistaken for Glaucous or even Iceland Gulls. Careful consideration of head shape and bill color and proportion should reveal the true identity.

Glaucous Gull ***Larus hyperboreus***

LENGTH 25–30 IN.; WING SPAN 56–60 IN.

Hyperboreus may be translated as "beyond the north wind," and these are truly Arctic creatures. A few individuals occur annually along the coast. Since most of these rarities are young birds in first-winter plumage—nearly pure white—they are easy to pick out from the darker gulls with which they mingle. The clinching field mark of these first-cycle birds is the pink-based bill with a dark tip sharply defined.

California Least Tern ***Sternula antillarum browni***

LENGTH 9 IN.; WING SPAN 20 IN.

This spry and acrobatic little tern, the smallest of the tribe, occurs only in the southern reaches of the north coast, occurring north of San Francisco Bay only as a wanderer. Important nesting colonies located in Alameda, Contra Costa, and Solano Counties are vulnerable to disturbance and depredation so are closely monitored each year.

In summer, Least Terns can be seen foraging close to shore along the eastern edge of San Francisco Bay, in buoyant flight, hovering and diving, especially over calm tidal areas as they hunt for small schooling fish. In winter, Least Terns move southward along the coast of Mexico.

SPECIAL STATUS Federally Endangered; state Endangered.

Caspian Tern *Hydroprogne caspia*

LENGTH 21 IN.; WING SPAN 50–55 IN.

Pl. 98

The largest of the terns, the Caspian Tern is common at estuaries and other coastal wetlands from mid-March to mid-September. The massive red bill is distinctive; its slow, powerful flight and the frequently given flight call—a hoarse "cwaaak"—are easily recognized. These boisterous calls are often heard overhead in spring and summer, even at night. Caspians are colonial, nesting in large numbers on remote levees and dry salt ponds and flats at protected (predator-free) places around the shore of San Francisco Bay (e.g., Brooks Island) and a few other sites along the coast: Humboldt Bay (Sand Island), Elkhorn Slough, and the Salinas River mouth. Because of their large size and aggressive nature, Caspian Terns can be a keystone species, defending territory from predatory gulls (and others) and inadvertently protecting smaller terns within the nesting colony.

Adults may fly long distances to forage for their chicks. Birds with nestlings on levees of South Bay are known to forage for fish off Santa Cruz in Monterey Bay. Birds foraging in Tomales Bay are thought to commute from the Napa River, a distance of over 30 mi in direct flight. They may make this long commute more than once a day when feeding chicks.

Dependent young appear with their parents on the outer coast in July and will follow them, begging constantly, to Mexico where they will spend winter.

Forster's Tern *Sterna forsteri*

LENGTH 13 IN.; WING SPAN 30 IN.

With only rare exceptions, any tern seen along the Northern California coast in winter is a Forster's. You can count on it and impress your friends with quick and distant identification—but first, be sure the bird isn't a similarly white and buoyant Bonaparte's Gull!

The Forster's Tern is the only tern restricted nearly entirely to North America year-round, present year-round north to Bodega Harbor and uncommon farther north as fall and winter visitors to Humboldt Bay and Crescent City. Primary nesting sites are in San Francisco Bay; elsewhere (e.g., Moss Landing) nesting occurs sporadically.

Elegant Tern *Thalasseus elegans*

LENGTH 17 IN.; WING SPAN 34 IN.

Along with Brown Pelicans and Heermann's Gulls, Elegant Terns form "the three amigos." While they may not really be friends, they do have a lot in common. All three nest to the south—mostly on islands in the Gulf of California—then migrate north to spend four or five months banqueting on fish in the cool, nutrient-rich California Current. Elegant Terns are following the Northern Anchovy *(Engraulis mordax)*, and their dispersal is correlated to the changing oceanic conditions (El Niño–Southern Oscillation) that drive anchovy numbers. From late June into October, Elegant Terns are common on the outer coast, surging in El Niño years, gathering in flocks of thousands at Bolinas Lagoon, Marin County, and in hundreds elsewhere, for example the Pajaro River mouth, Monterey County. A social species, this tern roosts in dense flocks on sandbars and beaches when traveling. Postnesting dispersal takes them into Oregon, but numbers decrease northward and the species heads southward in winter, leaving California behind. In warm-water years Elegant Terns move farther north and stay north longer than in cool-water years.

Gallinaceous Birds

Families Odontophoridae and Phasianidae

These chicken-like birds are among the most familiar birds to the general public. Only two species are common in coastal habitats: the native California Quail *(Callipepla californica)* (pl. 99), the state bird; and the nonnative Wild Turkey *(Meleagris gallopavo)*. The Ruffed Grouse *(Bonasa umbellus)*, in addition, approaches the coast in the coniferous forests of the northernmost counties.

The turkey, successfully introduced as a hunting species by California Department of Fish and Game in the 1970s, has become a nuisance in many coastal habitats, displacing quail and grassland birds, limiting oak reproduction, and disrupting the soil fauna as it forages in large flocks. The California Quail is a common bird of the coastal scrub, but it also inhabits chaparral, sagebrush, and grassland. Like other gallinaceous birds, quail gather in flocks after the nesting season, and it is not unusual to see dozens scurrying across an opening or foraging calmly in an open area, though rarely far from cover. Usually a sentinel will be perched above, watching over the flock and "clucking" as danger approaches. The three-note call of the male—"Chi-ca-go"—is a distinctive and familiar ingredient of the coastal California soundscape. Sadly, quail are nearly extirpated from San Francisco's Golden Gate Park because of predation by feral cats; yet happily they are still common in coastal shrublands.

One curious behavior exhibited by California Quail is dust bathing. Apparently this behavior is aimed at feather maintenance. The sand or dust is thought to absorb moisture and grease from the feathers, preventing matting and helping to maintain the insulating qualities of the plumage, as well as aid in removing dried skin, remnant feather sheaths, and ectoparasites.

Like the California poppy and the coast redwood, the California Quail is emblematic of the Golden State.

Pigeons and Doves

Family Columbidae

The common names of the Columbidae are somewhat interchangeable, but in general, the smaller species are referred to as "doves," the larger species as "pigeons." Family characteristics include small head, short bill, short legs, and a relatively chunky body. Large flight muscles in the breast account for this chunkiness. In general, they nest above the ground in trees, cliffs, or man-made structures. Although the family as a whole (more than 300 species) includes many colorfully plumaged birds, the North American varieties are rather plain colored, though with distinctive markings on the hindneck, in some cases iridescent. Members of the family are either seed-eating or fruit-eating species. Vocalizations are varied but tend to be low-frequency cooings or soft, owl-like hoots, often in paired phrases or notes. Ten species occur in California, four of which are common along the north coast; two are native species and two are exotic. The native Band-tailed Pigeon *(Patagioenas fasciata)* is a flocking, fruit-eating forest bird, often associated with elderberry bushes. The native Mourning Dove *(Zenaida macroura)* (pl. 100) is a seed-eating bird of open country, usually seen perched on a wire, foraging on the ground, or in strong, direct flight. The two nonnatives are more associated with urban environments. The Rock Pigeon *(Columba livia)* is the familiar "city pigeon" or "rock dove" known to all. The Eurasian Collared Dove *(Streptopelia decaocto)* is a relatively recent arrival to North America (introduced to the Bahamas in the mid-1970s) but is quickly spreading across the continent and has been colonizing the north coast since about 2005. Collared Doves seem to be displacing native Mourning Doves, at least in more urban environments, but this is a recent development not well studied or well documented.

Owls

Families Tytonidae and Strigidae

Think of the shape of an owl's head as a parabolic reflector or a satellite dish, designed to capture sound. The elliptical or round face has feathering arranged and specialized to direct sound waves to the ear holes. The ears, orifices in the "cheeks," are offset from one another, asymmetrical, allowing this predator to triangulate on the subtle sounds of a mouse scurrying through the grass, or of a tree vole climbing through spruce needles. (The "ear tufts" of some species are not hearing organs, rather expressive display feathers.) The eyes too are designed for visual acuity—disproportionally large, able to see in extremely low-light conditions, but fixed in place, surrounded by a bony structure called the sclerotic ring. The fixed eyes require owls to swivel the head in various directions to widen their peripheral vision, movement that also aids in triangulation on sound. As a rule of thumb, owls with yellow eyes (like the Great Horned Owl) are more crepuscular or diurnal, and those with black eyes (like the Spotted Owl) are more nocturnal. Owl talons are exceptionally large and powerful, designed for gripping and crushing prey. The beak is short, falcate, and sharp—a tool for tearing flesh. Flight feathers have fringe-like, serrated leading edges ("flutings") that reduce air turbulence and allow nearly silent flight, a sneak attack on prey. Plumage is generally various shades of browns, grays, and buffs—cryptic coloration designed for stealth and camouflage. As in diurnal raptors, reverse sexual size dimorphism is typical of owls, with females larger than males in body size; this is especially true of larger owl species.

Fourteen species have been recorded in California (including the Burrowing Owl, pl. 101), though only six can be considered relatively common along the north coast—three large, three small. Large, bold, and crepuscular, the Great-horned Owl *(Bubo virginianus)* is the most familiar of the owls. The Barn Owl *(Tyto alba)* is also common and familiar, a light-colored bird of open country. The Northern Spotted Owl *(Strix occidentalis caurina)* is probably the most renowned because of its threatened status and an association with old-growth forests that are coveted by the logging industry. The Barred Owl *(Strix varia)*, a relatively recent immigrant from the east, is causing some serious concern about its effect on its Spotted cousin—displacement, predation, and hybridization.

Three smaller species also occur in coastal woodlands. The Western Screech-Owl *(Megascops kennicottii)* is a bird of oak woodlands. The Northern Saw-whet Owl *(Aegolius acadicus)* is associated primarily with moist north-coast conifer forests and riparian lowlands. The Northern Pygmy-Owl *(Glaucidium gnoma)* is found in both dense evergreen forest and drier, more open mixed-evergreen forest, often near the forest edge.

Seasonal movements are not well understood, but the Saw-whet seems to be the most migratory, and even irruptive in some years. The Screech-Owl is the most sedentary (non-migratory) of the three. The Pygmy-Owl is not migratory but may move to lower elevations or into marginal habitats in the nonbreeding season. Of the three smaller owls, the Pygmy-Owl is perhaps the least coastal.

"Northern" Spotted Owl *Strix occidentalis caurina*

LENGTH 18 IN.

Pl. 102

This most coastal of three subspecies is a nonmigratory, year-round resident through each coastal county from Marin to Del Norte, and northward into southwestern British Columbia. This is a forest-core species, and the structure of the forest—multilayered, closed canopy with some dead standing trees and a diverse understory—determines its suitability for this owl. These are nocturnal predators specializing in medium-sized rodents: in the southern portion of its coastal range, Dusky-footed Woodrats *(Neotoma fuscipes)* are the preferred prey; to the north, Northern Flying Squirrels *(Glaucomys sabrinus)* provide the primary prey base. Spotted Owls have a distinctive vocal repertoire consisting of loud hoots, barks, and whistles. As is true in other medium to large owls, males are slightly smaller than females, but they have noticeably lower-pitched calls. Spotted Owls form long-term pair bonds and often roost together under the canopy in daylight hours throughout the year.

SPECIAL STATUS Federally Threatened.

Barred Owl *Strix varia*

LENGTH 21 IN.

Very similar in appearance to the Spotted Owl, the Barred Owl inhabits a broader range of habitat types and is slightly larger and more aggressive than the Spotted Owl. Formerly confined to forests east of the Mississippi, over the past century or so the Barred Owl has expanded its range westward, bringing it into contact with its gentler cousin over the last three decades. The reasons for this expansion are not clear, but human modification of the landscape—extensive logging of boreal forests, fire suppression, and restoration of riparian forest corridors in the northern Great Plains—have been suggested and suspected as factors removing natural barriers to dispersal. Whatever the cause(s), Barred Owls are now displacing Spotted Owls, driving them from formerly occupied territory, competing for food, and even killing or hybridizing with them. A controversial new plan to control (depredate) Barred Owls where they threaten Spotted Owls has been proposed by the U.S. Fish and Wildlife Service (USFWS 2012).

There is a conservation conundrum here, in that the barred owl is a native species that has expanded its range westwards, either naturally or with a degree of human facilitation, and now constitutes a major threat to the viability of another native species, the threatened spotted owl.

—R. J. Gutierrez et al. 2007, "The Invasion of Barred Owls and Its Potential Effect on the Spotted Owl"

Swifts

Family Apodidae

The family name of the swifts literally means "without feet," which is understandable given these birds' aerial habits—constantly on the wing, siphoning insects from the atmosphere, their only food choice. Although superficially like swallows, swifts are from a different and distant evolutionary branch, more closely related to the nighthawks and hummingbirds! (Indeed, these three families are among the few groups of birds able to enter a state of torpor to conserve energy.) Close observation reveals that swifts have an entirely different flight pattern than swallows do. Although a swift's tail is forked or cleft as in some swallows, and its wings are long and bowed back. The wing profile is formed by the flight feathers: elongated outer feathers (primaries) and shortened inner feathers (secondaries). Swifts fly with rapid, rhythmic wing beats, then glide on locked wings in a swept-back bow or sickle shape, often banking or rolling. The bill is tiny, but with a wide gape, and the feet are small, though well adapted for clinging to vertical surfaces like cliffs, chimneys, or hollow trees where they roost, especially during migration. Three swift species are regular in our region, though none particularly common. The Black Swift *(Cypseloides niger)* (pl. 103) is the largest and rarest, a high-atmosphere cruiser (sometimes called the "cloud swift"). It has a very limited (and decreasing?) nesting distribution along coastal cliffs behind waterfalls in Santa Cruz and possibly Monterey (Big Sur) Counties, though these sites may have been abandoned in recent years. The Vaux's Swift *(Chaetura vauxi)* nests in the Pacific Northwest, southward to Santa Cruz County, mostly within the coast redwood *(Sequoia sempervirens)* zone. Large numbers pass in autumn's southward migration, congregating at night to roost in traditional sites in Sonoma and Marin Counties. The White-throated Swift *(Aeronautes saxatalis)* is a year-round resident in our region, nesting along the coast from Humboldt County southward. The easiest to identify of the group, given its striking black-and-white plumage pattern, it is a gregarious bird—nesting, roosting, and foraging in sizable flocks, soaring deftly about like the "sky pilot" it's genus name implies.

SPECIAL STATUS Two species are considered Bird Species of Special Concern (breeding)—Black Swift, priority 3, and Vaux's Swift, priority 2.

Hummingbirds

Family Trochilidae

This New World family evolved in the New World tropics and radiated northward to grace our gardens. Hummingbirds are sexually dimorphic with polygynous males displaying brilliant colors, including iridescent throat and forehead markings; the females are more modestly dressed, though back feathers can also sparkle with an iridescent glow. Hummingbirds specialize in extracting nectar from flowers—hence the long, thin bill—but also hawk for small flying insects. They have very short legs, and though able to perch, they are unable to walk. They make up for this liability with incredible agility on the wing. They are able to hover in place while feeding, and to fly backward, a unique talent among the North American avifauna. In aerial breeding display, wing beats may be as fast as 200 beats per second. Vocalizations tend to sound insect-like to the human ear, a series of ticks, buzzes, and high-pitched chatter, yet each species' voice provides a distinctive clue to identification.

Of 14 species recorded in California (many rare), only two occur regularly in our region: the year-round resident and larger Anna's Hummingbird *(Calypte anna)* and the smaller migratory Allen's Hummingbird *(Selasphorus sasin)* (pl. 104). A third, the migratory Rufous Hummingbird *(Selasphorus rufus)*, nests along the extreme northwestern coast and passes through the more southern counties in numbers during migration, especially in autumn, following the departure of Allen's Hummingbirds.

Kingfishers

Family Alcedinidae

Nested in the order Coraciiformes—largely an Old World and tropical group that also includes motmots, todies, and bee-eaters—this family has complex taxonomic relations; however, we have only one species in California. All kingfishers are sit-and-wait predators with large heads and short necks, large eyes, and long, strong beaks. Our Belted Kingfisher is a fish specialist, though despite the name *kingfishers*, not all species (90 plus worldwide) eat fish. The feet are small and the three front toes are joined for part of their length (syndactyl). The body is compact, the wings are relatively short and rounded, and the tail is stubby.

The Belted Kingfisher, with regal crown and jaunty profile, often perches above calm water while foraging. Indigestible portions of its prey are regurgitated as pellets. *K.H.*

Belted Kingfisher *Megaceryle alcyon*

LENGTH 12–14 IN.

Pl. 133

Jaunty is a word often used to describe the spectacular Belted Kingfisher, probably because of the way he carries his large, crested head. The kingfisher is a year-round citizen of the coast at most aquatic habitats—intertidal, estuarine, riverbank, stream, and pond. When foraging, kingfishers sit and wait, then hover and plunge dive for surface fish and sometimes pollywogs, crayfish, or rarely newts and salamanders. At some places (such as Bolinas Lagoon, Marin Co.) several kingfishers sit on telephone lines along the shore, watching intently for shiny scales in the water. Each bird defends foraging territory aggressively, so they are usually perched at some distance from one another.

For housing, a kingfisher digs a tunnel in the face of a riverbank or eroded cliff. It can be 8 ft deep, with an enlarged nest cavity at the end. During the four weeks from hatching to fledging, the young are fed by their parents in virtual darkness. On the day of fledging, their first launch into *flight* is also their first launch into *light!*

Kingfishers are rather vocal birds, but their repertoire is limited. Most commonly heard is a loud rattling call, reminiscent of a woodpecker (to which they are closely related), and usually uttered in flight; this is often the first clue to a kingfisher's presence. Both sexes are blue backed and have a broad band across the upper breast, but females ("queenfishers") have a deep rusty wash along the flanks; males are white underneath.

Woodpeckers

Family Picidae

This is a diverse family with 17 species in California; six are rather common in coastal forests. The woodpeckers are well named, as the primary goal of most species is extracting insects from rotting trees, but some also eat nuts, fruit, and sap. The family is well adapted for climbing tree trunks—most species have X-shaped (zygodactyl) toes, stiff tails feathers, and highly sensitive scent organs for finding food. For extracting food, they are armed with a stout chisel-like bill and a sticky long tongue. The skull is reinforced, helmetlike, to endure constant head banging against hard surfaces. Although their vocal repertoire is limited to rattles and cackles, most species drum on wood, producing loud territorial noises. All are hole nesters and, as such, are important keystone species within their communities, providing nest sites in trees for secondary nesters such as small owls, swallows, wrens, and nuthatches. Woodpeckers have a characteristic flight pattern—tracing a flap-and-glide undulating pattern across the sky. The social system within the family is variable: most are solitary nesters; others such as the Acorn Woodpecker *(Melanerpes formicivorus)* (pl. 105) form colonies of cooperative family groups. Most species show some degree of plumage dimorphism, with males having more red on the head and nape than females. Besides the Acorn Woodpecker of the oak woodlands (most coastal in the southern portions of our region), other regular species include the Red-breasted Sapsucker *(Sphyrapicus ruber)*, the similarly dressed Downy and Hairy Woodpeckers *(Picoides pubescens* and *P. villosus)*, the Pileated Woodpecker *(Dryocopus pileatus)*, and the rather anomalous Northern ("Red-shafted") Flicker *(Colaptes auratus)*. Unlike the others, the flicker is primarily a ground-foraging species that digs for ants and other insects.

Falcons

Family Falconidae

Different models of diurnal raptor than the hawks and their allies, the falcons have streamlined bodies but with proportionally deep chests; long, tapered wings and tails; and large feet with long, sharp talons on relatively short legs. The talons, the primary weapons of the falcons, are rather large and have an elongated middle toe and an especially powerful hind toe. The falcons are extraordinary flying machines, designed by evolution to outfly and outmaneuver the most aerobatic of prey or the most aggressive opponent. Only three species are regular on the coast—American Kestrel, Merlin, and Peregrine (in order of both taxonomy and increasing size)—and each occupies a refined niche among the birds of prey. Sexual dimorphism is more pronounced than in other raptors: females are larger than males with little overlap, and plumages differ between sexes to varying degrees, but this is most pronounced in the diminutive American Kestrel.

In the 2012 supplement to the Checklist of North American Birds (AOU), the phylogenetic position of the falcons was revised, placing them between the woodpeckers and the songbirds, a radical shift from their former position next to the eagles and hawks, earlier in evolutionary sequence.

American Kestrel ***Falco sparverius***

LENGTH 8.5–10.5 IN.; WING SPAN 20–24 IN.

Pl. 106

Kestrels, formerly called Sparrow Hawks, are small, dainty falcons that hover and drift when searching for prey. Females are noticeably larger than males. Both are foxy brown barred with black on the back, but males have blue-gray wings and are among the handsomest of our raptors. Both sexes have vertical black bars on the cheeks, one behind and one in front of the eye. Two black spots on the back of the head are thought to be false eyes, perhaps used to frighten larger predators. Indeed, when approached by larger raptors, kestrels often swivel their heads around to "face" the intruders with the back of the head. Most of their diet is composed of large insects (beetles and grasshoppers), and they will readily take earthworms, lizards, and small mammals (rodents and shrews), but rarely small birds, thus debunking the still-used nickname "sparrow hawk."

Kestrels are most often seen perched on fence posts or telephone wires from which they watch for prey, which they dart down to capture on the ground. Foraging habitat is short-grass pasture or barren ground. In our region, kestrels are more common in winter than in summer.

Kestrels are secondary hole nesters, occupying old woodpecker and flicker excavations or naturally occurring cavities, even the eaves of houses or nest boxes. They can be quite abiding and tolerant of human presence, foraging in people's yards from rooftops and garden fences. Numbers increase in winter, especially in the more southerly counties, with an influx of birds from less hospitable climes.

Merlin *Falco columbarius*

LENGTH 9.5–13.5 IN.; WING SPAN 21–27 IN.

Pl. 107

This is among the most energetic and beguiling of all the raptors, and a "Merlin day" provides a rush of jubilance for every enthusiastic naturalist. Only slightly larger than the kestrel, dashing and self-assured, the Merlin is an avian magician. When hunting small shorebirds, the Merlin seems to appear out of nowhere, strafing low and bullet straight across the tidal plain. While kestrels drift and float like overgrown swallows, Merlins race and chase with deep and decisive wing beats at warp speed. Even when perched, the Merlin constantly looks all around and its head bobbles up and down, always ready for the next predatory opportunity.

Merlins are primarily bird hunters, and along the outer coasts many specialize in killing Dunlin and Western Sandpipers. Often, the first clue that a Merlin is present is the panic-stricken behavior of the local "peeps," nervous clouds of them swirling over the tidal flats in evasive maneuvers.

Merlins do not nest in California but arrive in fall and leave in spring, their passage coinciding with the movement of wintering shorebird flocks. The underparts are heavily streaked, and the browner females are considerably larger than the bluish males. Of three distinct North American subspecies, the "boreal" Merlin of the northern forests *(F. c. columbarius)* is most often seen in our region.

Peregrine Falcon *Falco peregrinus*

LENGTH 14–23 IN.; WING SPAN 37–43 IN.

Pl. 108

Powerful, fast, and elegant, the elite athlete among the raptors, a Peregrine causes widespread panic among all smaller coastal birds, just by flying past. Sexes are best separated by size, with the female 15 to 20 percent larger than the male and up to 50 percent bulkier (heavier). Size dimorphism in Peregrines, as in many other raptor species (Osprey, accipiters, some owls), allows a pair to capture a wider range of prey sizes. Some pairs nest in remote localities of Northern California, but around San Francisco Bay, Peregrines nest on bridges, building ledges, and power towers. Once a rarity due to persecution and environmental contaminants (DDT), Peregrines have made a dramatic comeback aided by protections under the Endangered Species Act and an aggressive captive-breeding program under the direction of the Santa Cruz Predatory Bird Research Group. The success of those efforts resulted in the removal of the Peregrine from the federal endangered species list in 1999. Now Peregrines are seen on almost every visit to the larger bays and estuaries or coastal headlands, any place where there are rafts of ducks to flush or shorebirds or seabirds to harass.

The species is most common from October through March, when birds are visiting from Alaska and Canada. Look for Peregrines perched on power poles or snags near coastal wetlands, or on cliff faces at rocky headlands. Their presence is often betrayed by the frantic screams of Willets and terns and flushed shorebird flocks as the Peregrines strafe tidal flats or the shores of lagoons and estuaries.

Three races occur in California, each with unique plumage characteristics, but the most common by far along the north coast is the locally nesting race, *P. f. anatum*. The similar Prairie Falcon *(F. mexicanus)* is a rare visitor to the coast, much lighter on the dorsum and with black "wingpits" (axillaries).

SPECIAL STATUS Formerly federally Endangered, now delisted.

Perching Birds (Passerines)

Family Tyrannidae

The North American tyrant flycatchers prey primarily on flying insects. Typically they perch in an opening or atop the vegetation and sally out to snap up a flying insect, often returning to the same perch. Tail wagging is common in the family and may provide a useful clue to identification. The bill is broad and flattened. Most family members are migratory,

PASSERINES

The passerines are smallish birds with feet and leg muscles adapted for perching, hence another name: perching birds. Chicks hatch from the egg featherless (altricial) and are dependent on parental care for the first few weeks of life while they develop in the nest. The passerines are divided into two major groups (or clades): the suboscines and the oscines. The suboscines are the more primitive of the two and include the tyrant flycatchers (Tyrannidae). They do not have the well-developed vocal musculature of the true songbirds. Their vocalizations are innate (not learned) and tend to be less complex and less musical than those of the true songbirds, the oscines.

but the Black Phoebe *(Sayornis nigricans)* is resident along the Northern California coast, although northernmost populations may shift south in the winter. Black Phoebes are a reliable presence anywhere near water. The similar Say's Phoebe *(S. saya)* is a bird of open country and nests in the interior but moves coastward in winter, appearing along our more southerly shores from September into April. Say's Phoebes are often seen on tidal marshes, in short-grass pastureland, or along the outer beach. The Pacific-slope Flycatcher *(Empidonax difficilis)* (pl. 109), the smallest of these, is a woodland bird and one of the common breeders in humid coastal woodlands—riparian, mixed broadleaf-evergreen, and conifer forests. It migrates southward to winter in Mexico's lowlands of the Pacific coast. The Western Wood-Pewee *(Contopus sordidulus)* is a drab forest dweller, also migratory, also common, and rather difficult to identify except by call, a nasal or burry, "pee-eeer." Another, larger member of the tribe, the Olive-sided Flycatcher *(Contopus cooperi)*, also a summer resident, breeds in coastal and "near-coastal" conifer forests through the length of our region. Its distinctive three-note call, "Quick, three beers," is usually given from a high perch.

SPECIAL STATUS The Olive-sided Flycatcher is a Bird Species of Special Concern, priority 2.

SONGBIRDS

In songbirds, or "oscines" (from the Latin word for the same), the musculature controlling their song box (syrinx) is much more complex than that of more primitive species (the suboscines), therefore most songbirds have more complex, elaborate, and, to the human ear, beautiful vocalizations than those that precede them in taxonomic order. Though born with rudimentary vocal ability, oscine vocalizations are refined by the songs of their parents, and neighbors, during the first several weeks of life.

> The earth has music for those who listen.
>
> —William Shakespeare

Family Vireonidae

Although superficially similar to wood warblers in size and behavior, the vireos are from a different evolutionary line. The relationship of the vireos to other New World songbird families is uncertain, but it is thought that the group arose from a common Old World ancestor (monophyletic). The AOU Checklist states: "There seems to be no evidence for their placement within the nine-primaried oscines," the group that contains the rest of North America's songbird families (see above). Vireos tend to be slightly stockier than wood warblers and have thicker bills with a noticeable hook at the tip. The vireos are leaf gleaners, though some may "flycatch" on occasion. Each species sings tirelessly, sometimes even from the nest. Sexes are alike, but females do most of the nest attendance and males most of the singing.

Of the 12 species that occur in California, three are regular nesters along the immediate coast, all in the genus *Vireo*: Hutton's Vireo *(V. huttoni)* (pl. 110), Cassin's Vireo *(V. cassinii)*, and Warbling Vireo *(V. gilvus)*. The Hutton's is associated mostly with oak woodlands and is the only year-round resident in the group. The Cassin's, a neotropical migrant, occupies mixed oak-conifer and deciduous riparian forest along some of its coastal range. The Warbling Vireo, also a neotropical migrant, is allied with mixed deciduous woodlands, on both the edge and interior of the forest. As its name suggests, the Warbling Vireo has a melodious, liquid song of continuous notes; like others of the group, it is very vocal in the spring. The Cassin's Vireo's song is a series of "burry," punctuated, and somewhat querulous phrases. (Beware: the Purple Finch has a vireo-like song that is easily confused with the Cassin's Vireo.) In contrast, the Hutton's Vireo has a raspy, much less complex (but equally persistent) voice, a series of unmusical "zu-wee, zu-wee . . ." repeated over and over as the bird moves through the foliage.

Family Corvidae

The corvids—crows, ravens, and jays—with over 120 species worldwide, are prominent members of their communities throughout their range. They are the most intelligent and perspicacious avian family, a trait that endowed the Common Raven with demigod status in some aboriginal American cultures. Physically corvids are sturdy birds—stout-billed, solidly built, and rather large. In fact, the family includes the largest of the passerines. Sexes look alike. Behaviorally, corvids tend to have highly developed social organization and strong pair bonds, and they display exceptional intelligence, or survival skills. Most species exploit a wide variety of resources and frequent human-dominated environments. They are notorious nest robbers—devouring eggs and even nestlings.

Common Raven ***Corvus corax***

LENGTH 24–30 IN.

Pl. 111

Common Ravens are obvious residents all along the coast from San Mateo County north. Because raven populations are increasing throughout the western states, the species has recently become more widespread in Monterey and Santa Cruz Counties as well.

Extraordinarily astute, these largest corvids *rule* the beaches within their territory. They seem to know exactly when a dead crab washes up, where the Gray Fox spends its day, when human spillage will be abundant (weekends), and where and when each Snowy Plover lays her eggs. Ravens are omnivores with a foraging strategy that includes predation, and their diet is impressively eclectic—intertidal invertebrates, terrestrial insects, rodents, shrews, small land birds, eggs and young of Snowy Plovers, and any carrion that washes up along the shore, from a mole crab to a beached whale.

The population explosion of Common Ravens in coastal California (as well as throughout the state) has followed that of humans, subsidized by landfills and road-killed wildlife, on which the ravens feed. Through the mid-1900s ravens were rare in San Francisco and other urban centers; by the year 2000 they had become common. In 2010, they first nested on the Farallon Islands, 20 miles offshore, a worrisome event in the minds of the islands' seabird biologists, given ravens' predatory proclivities. Because they are such effective nest and fledgling predators, ravens have been implicated in the decline of the Marbled Murrelet population, especially in Central California where the murrelets' nesting colonies are located near campgrounds that attract scavenging ravens. The Western Snowy Plover has also suffered from raven depredation on coastal beaches, and the National Park Service has made a valiant effort to protect nesting plovers with constructed exclosures, trash control, and plover protection zones.

Those who dream by day are cognizant of many things that escape those who only dream by night.
—Edgar Allan Poe, "The Raven"

American Crow ***Corvus brachyrhynchos***

LENGTH 16–20 IN.

American Crows are common along the coast where agriculture, pastures, or human development are close to the beach, but they are absent at wilder, more remote headlands.

American Crows are smaller than Common Ravens in all dimensions including attitude. From below, their tails are shorter and squared at the tip, not wedge shaped like the Common Raven's tail. Up close, a crow's bill is smaller and less deep than a raven's, and the head and neck feathers of a crow are rather smooth; a raven's crown feathers are spikier and the neck feathering mane-like. Crows tend to gather in substantial

Separating the raven (top) and the crow (bottom) is one of the first challenges facing novice naturalists. This drawing shows the relative proportions of the two. Note the difference in bill size, the length of the wings relative to the body, the bulk of the body, and most diagnostically, the shape of the tail. *K.H.*

flocks; ravens are usually seen in pairs. And the flight patterns of the species differ: crows tend to row across the sky, with strong and steady wing beats. Ravens, on the other hand, are quite aerobatic, wheeling and diving, gliding and sometimes rolling across the skyscape, as if showing off. So, crows don't soar—ravens often do.

Family Alaudidae

This is an Old World family with only one North American representative. Larks are birds of dry, open habitats; they nest and forage on the ground. Most have complex songs and sing in flight. The hind claw (hallux) is longer than on most passerines, an adaptation for walking on the ground.

Horned Lark *Eremophila alpestris*

LENGTH 6 IN.

Plowed fields, short-grass prairie, and dunes are home to the ground-dwelling Horned Lark. In our more southerly counties, small numbers nest along the coast, most often in the hind dunes of the coastal strand. Many more come from the north and east to join the locals in winter. Larks are gregarious and often flock with American Pipits *(Anthus rubescens)* in the nonbreeding season and wander across open country in search of food (seeds and insects). Like pipits, the Horned Lark has white outer tail feathers, a trait common in several flocking species. This species is wary of mammals—flying away at the slightest hint of danger, so any good view of its strikingly patterned face painted yellow and black and the curious "ear tufts" is a gift. The high-pitched, chittering song is often given in flight or from a low perch near the nest site.

The same species is called Shore Lark in the United Kingdom.

SPECIAL STATUS California Department of Fish and Game Watch List.

Family Hirundinidae

Swallows and martins are streamlined, agile fliers, with long, narrow wings, a short beak with a wide gape, short legs, and small feet, all adaptations for aerial foraging.

Of eight species recorded in California, seven occur in our region. Of those, five are rather common, two rather rare. The common ones—the Tree Swallow *(Tachycineta bicolor)* (pl. 112), Violet-green Swallow *(T. thalassina)*, Northern Rough-winged Swallow *(Stelgidopteryx serripennis)*, Cliff Swallow *(Petrochelidon pyrrhonota)*, and Barn Swallow *(Hirundo rustica)*—are widely distributed through the region. The rarer species—the Purple Martin *(Progne subis)* and Bank Swallow *(Riparia riparia)*—are much more localized in distribution, and both are considered species at risk. Martins are hole nesters and compete with European Starlings for nest sites.

SPECIAL STATUS The Purple Martin is a Bird Species of Special Concern, priority 2.

Bank Swallow *Riparia riparia*

LENGTH 5.25 IN.

The Bank Swallow is a wide-ranging species, with a breeding distribution that is largely holarctic; its wintering distribution is confined mostly to the Southern Hemisphere. Bank Swallows formerly nested the length of the California coast but are now extirpated from Southern California because of coastal development and river channelization. Most of the remaining nesting sites are in the interior lowlands (e.g., Sacramento and Feather Rivers), with a few protected colonies remaining scattered along the north coast—Smith River (Del Norte Co.), Fort Funston (San Francisco), Año Nuevo (Santa Clara Co.), and Pajaro River (Monterey Co.). Bank Swallows are highly colonial, and as the name implies, they nest in riverbanks or coastal bluffs where the soil is friable enough for them to excavate burrows. The nest sites are rather shallow, extending about 6 to 36 in. into the bank. Wherever they nest, flooding, bank erosion, human disturbance, and predation by snakes and birds pose challenges.

Bank Swallows arrive at their traditional breeding sites in spring. After the young have fledged (18 to 22 days after hatching), the birds remain in the general area of the colony for only a few weeks before dispersing. Most are gone by midsummer. The most reliable time and place to see Bank Swallows are late spring to early summer as the birds forage over still water near the nesting colony.

SPECIAL STATUS Listed as Threatened by the California Department of Fish and Wildlife.

Families Paridae and Aegithalidae

In Europe, the common name *tits* is given to the family of chickadees and titmice, and to the family of bushtits, because of their small size. These are plump, active, and gregarious birds of forested environments. The small, pointy bill of the chickadees is adapted for gleaning insects from foliage, but they also visit seed feeders readily. Titmice eat insects and seeds, and also fruit, but their blunter bill may be used for chipping away bark in search of larvae. All the North American species of the Paridae are secondary hole nesters, meaning that they use old woodpecker holes, natural crevices, or nest boxes, but they may also excavate their own nests in rotting wood. Only three species occur in our region. The Black-capped Chickadee *(Poecile atricapillus)* is more common on the north coast, from Humboldt County north, a bird of the boreal conifer forest. The Chestnut-backed Chickadee *(Poecile rufescens)* (pl. 113) is associated with moist conifer forest, especially in the northern counties, and with mixed oak woodlands, riparian habitats, and exotic plantings in suburban areas in the more southerly counties. Black-capped and Chestnut-backed overlap in the northernmost counties, where Black-capped are restricted mostly to hardwood forests and riparian corridors. The Oak Titmouse *(Baeolophus inornatus)* is a bird of dry oak woodlands, occurring in the more southerly portion of our region, absent from the humid north coast. In the nonbreeding season, titmice may disperse coastward, especially along riparian corridors.

Bushtit *(Psaltriparus minimus)* is considered a close relative of the Paridae in some classifications, so the two groups are "lumped" together here. These acrobats of the bushes are highly social birds and are usually encountered in sizable flocks. The Bushtit is common in various coastal environments, especially brushy habitats associated with coastal scrub, and in riparian areas, chaparral, and forest edges. Long tailed and short winged, Bushtit is not a long-distance flier, rather a year-round resident throughout its range. Its hanging, socklike nest—elaborately woven from lichen, mosses, spiderwebs, and grasses—is one of nature's architectural wonders, designed to insulate this diminutive bird from the elements. Bushtits are known cooperative breeders, with several adults attending the eggs and young, although this behavior may be rare along the north coast.

Families Sittidae and Certhiidae

All the species in the nuthatch and creeper families are rather small-bodied tree-bark specialists that forage along trunks and pry or pick prey from beneath the bark or from the cones of conifers. Their feet, like those of woodpeckers, have strong opposable toes and curved claws, specialized for grasping and climbing vertical surfaces (scansorial). The bills of the species differ somewhat—curved in the creepers, stubbier in the nuthatches—each designed for slightly different methods of foraging. Creepers have long, stiff tails, whereas nuthatches are very short tailed.

The Red-breasted Nuthatch *(Sitta canadensis)* and Pygmy Nuthatch *(S. pygmaea)* are common members of coastal forests, the former most common in dense, mature conifer forests (redwood, Douglas fir), the latter in drier, more open pine forests (e.g., Monterey and bishop's pines). Red-breasted Nuthatches are irruptive, with populations exploding in some productive years and becoming much more common in coastal forests in the nonbreeding season. Pygmies (fondly referred to as "peanuts" by birders) are distributed through the more southerly counties of our region, mirroring the distribution of long-needled pines. The Pygmy Nuthatch is a cooperative breeder. The White-breasted Nuthatch *(S. carolinensis)* prefers deciduous hardwood forests dominated by oaks. This nuthatch is rare along the coast north of San Francisco Bay, less so to the south.

The "Pacific" Brown Creeper *(Certhia americana occidentalis)* is a common but often overlooked bird of coastal forests. It is the only member of its family in North America, although there is some evidence, based on different songs and mDNA analysis, that several different species may be hidden within the complex of subspecies found on the continent. This phenomenon, known as "cryptic species complex," was exemplified recently by the splitting of the "Winter Wren" into three separate species (see Family Troglodytidae). Although essentially indistinguishable in the field, variations in song among regional populations may reveal the presence of separate species. In the case of the creepers, this geographic boundary appears to be located at about 32°N latitude, near the southern border with Mexico.

Family Troglodytidae

Wrens are small birds, most often found close to the ground, with highly developed vocalizations. Most are accomplished songsters with elaborate vocal repertoires. Wrens may have dialects that vary across small distances. Sexes are alike in plumage, and most species have a brown or rufous ground color with dark barring on the wing and tail feathers. Five species occur regularly in our region. Pacific Wren *(Troglodytes pacificus)* (pl. 114), formerly in the Winter Wren complex, is a common member of the coastal conifer community where its bubbling song echoes through coastal canyons and forests. Bewick's Wren *(Thryomanes bewickii)* is a common and vocal resident in coastal scrub, chaparral, forest edges, and suburban environments. House Wren *(Troglodytes aedon)* is present in the southern counties near the coast, but northward it occurs only at the drier, interior sites, with some individuals moving coastward in autumn. Rock Wren *(Salpinctes obsoletus)* is generally less common than the others, favoring rocky headlands, outcroppings, and quarries along the coast. Some ornithologists suggest that there are two species of Marsh Wren *(Cistothorus palustris)* in North America—the Eastern and Western.

Marsh Wren ***Cistothorus palustris***

LENGTH 4–5 IN.

"Sassy" may best describe the character of this marsh denizen. Seldom seen except when singing from the tops of bulrushes or cattails in spring, Marsh Wrens "gargle, twitter, and chatter" from most sizable fresh or brackish marshes all along the coast. Indeed, it may seem that these voluble wrens are competing with Red-winged Blackbirds as the noisiest fellows in the neighborhood. The males are exceptionally vocal and learn an incredible diversity of song types early in life. Although most active during daylight hours, Marsh Wrens may also sing at night, especially under the light of the moon.

Only about 60 percent the size of a Song Sparrow, the Marsh Wren is the smallest of the "little brown jobs" in marshy habitats. Some male Marsh Wrens build numerous nests within their territory each year, a necessary complement to their polygynous mating strategy (a behavior more common in the west than the east). One nest houses the female, the eggs, and the young, while the male roosts in another. He also builds several "dummy" nests that stay empty all season. This behavior may have evolved as predator avoidance—a raccoon loses interest after dismantling an empty nest—or maybe the construction just gives the Marsh Wren something to do with its time.

Marsh Wrens are insectivores, foraging low in the vegetation or on the marsh floor, primarily on insects and spiders.

Family Cinclidae

The five species of dippers, all in the same genus (therefore closely related), are distributed through the Americas, Asia, and Europe, although there is only one North American species. All dippers have plump bodies, short wings, and stout legs, and all are associated with swift-flowing streams. Because of the short wings, dippers have a distinctive flight pattern—rapid, shallow wing beats often interspersed with gliding—following a stream course low above the water. Physiological adaptations to their diving habits are common in waterfowl, but rare in passerines. In dippers, these include reduced metabolism (heart rate) when diving and a high concentration of hemoglobin in the blood to hold oxygen when underwater.

American Dipper — *Cinclus mexicanus*

LENGTH 7–8 IN.

Once called Water Ouzels—from the Old English meaning "blackbird" or "thrush"—dippers are landbirds that have adapted so well to water that the American Dipper is "North America's only truly aquatic songbird." It is the only passerine that routinely swims and dives underwater into churning rapids where it extracts small fish or larval insects from the creek bed. As Ralph Hoffman proclaimed in his classic *Birds of the Pacific States* (1927), it "is an astonishing sight to see a bird . . . dive headlong into a foaming mountain stream and disappear under its waters." I (R.S.) have seen a dipper pop out of the water with five small fish neatly stacked in its bill. How does she catch the fourth without losing the first three?

The American Dipper does not occur *on* the immediate coast but does live just upstream where wild rivers or streams with permanent surface water make it to the sea. It is most likely to be found in the more northern reaches of our region; sightings in coastal streams south of Mendocino County are noteworthy. Adaptations for its water-loving ways include an actively winking white eyelid (nictitating membrane), valve flaps over the nostrils, a thick waterproofed plumage, a habit of bobbing in place (as seen in other streamside species such as waterthrushes and Spotted Sandpipers), a short tail, a stout physique, and powerful wings and legs for foraging in swift currents. The sexes are similar.

The song, having evolved to carry its message above the river's roar, is one of nature's purest sounds, as expressed by California's preeminent naturalist, John Muir, in his *Mountains of California* (1894): "[Its] music is that of the streams refined and spiritualized. The deep booming notes of the falls are in it, the trills of the rapids, the gurgling of margin eddies, the low whispering of level reaches, and the sweet tinkle of separate drops oozing from the ends of mosses and falling into tranquil ponds."

Family Sylviidae

This family is commonly called the Old World warblers. Most members are small to medium in size, with uniform or dull-colored plumage, long tails and short wings, and a fairly fine bill (though thicker than in most New World warblers) with bristles at the base. The diet is mostly insectivorous, though berries are also taken.

Wrentit ***Chamaea fasciata***

LENGTH 5–6 IN.

Pl. 115

Definitely not a waterbird, the Wrentit inhabits coastal-scrub habitat from the ridgetop to the beach, where the male's "bouncing Ping-Pong ball" song rolls out over estuaries and bays. And unlike most other passerines, Wrentits sing all year-round. Males and females look alike, both adults with a bright yellow iris that gives a piercing stare to those lucky enough to catch a glimpse.

Coastal scrub and dry understory vegetation of the forest is their habitat, through which they skulk, only occasionally taking flight on short wings. The tail is extremely long, often cocked upward, and serves as a reliable rudder when navigating through the shrubby tangles. The Wrentit has been called the "most sedentary of North American birds." Wrentits were almost endemic to California and Oregon, occurring only on the coastal slope, with a distribution ranging from northern Baja California to the Columbia River, a natural barrier to such a sedentary species. However, land-use practices that have favored shrub over forested habitats over the last century have allowed the species to extend its range somewhat, possibly one bush at a time.

The Wrentit's taxonomic relationships have proved a puzzlement to ornithologists. The species was once assigned to the Old World babblers (family Timaliidae) but more recently placed within the Old World warblers (Sylviidae). To whatever family it belongs, the Wrentit is a truly unique member of the North American avifauna, the only member of its genus (sui generis).

Family Turdidae

This family, generically called the thrushes, also includes bluebirds, robins, and solitaires. The birds are smallish (Western Bluebird) to medium in size (American Robin), with a strong and straight bill, relatively large eyes, and rounded wings. Although thrushes are not highly dimorphic, males tend to be brighter than females and fledglings tend to be spotted. Many species forage on the ground and in leaf litter, though some, like the bluebird, are "perch-and-watch" insect predators.

The Western Bluebird *(Sialia mexicana)* is a gentle bird of open country, frequenting coastal strand, prairie, and pastureland. There are few sights more dazzling than a brilliant male Western Bluebird hovering against the cerulean clear California sky. Bluebirds are here year-round, often in extended family groups, but seasonal movements are erratic, with birds moving inland, away from the immediate coast, when winter storms threaten. The Swainson's Thrush *(Catharus ustulatus)* and Hermit Thrush *(C. guttatus)* are the two most common forest thrushes. These two species are similarly dressed, but easily distinguished by voice and behavior. The Swainson's is highly migratory, therefore present only in late spring into early fall. The Hermit overwinters, with numbers increasing in winter, especially in the more southerly counties. The American Robin *(Turdus migratorius)* needs no introduction and is present year-round; large flocks sometimes gather in fall and winter where berry crops are abundant. The Varied Thrush *(Ixoreus naevius)* (pl. 116) is rather common in the boreal forests of the north coast, moving southward in the winter. (Note: Newly fledged robins in their first plumage superficially resemble the Varied Thrush.)

Family Mimidae

Confined to the Americas, the "mimids"—mockingbirds, catbirds, and thrashers—are medium-sized, rather slender birds with strong, sometimes down-curved bills, long tails, and rounded wings. All have complex and variable songs. Of the 35 species, nine occur in California and only two are regular in this region—the Northern Mockingbird *(Mimus polyglottos)* (pl. 117) and the California Thrasher *(Toxostoma redivivum)*. The Northern Mockingbird is a frugivore, and its range has expanded northward and coastward with the suburbanization of California and the planting of ornamental shrubbery. A loquacious songster and accomplished mimic, it is familiar to many suburbanites because of its improvisational vocalizations. "Mockers" incorporate local sounds into their repertoire, be they other bird songs, squeaky doors, or back-up beeps of garbage trucks.

California Thrasher ***Toxostoma redivivum***

LENGTH 12 IN.

Endemic to California's coastal chaparral, this largest of the thrashers is resident from southern Humboldt County southward into Baja California. Although of plain brown plumage, with a long tail and short wings, it is distinguished by its curved bill. The only other largish brown bird in the same habitat might be a California Towhee *(Melozone crissalis)*, but it has a stubby, sparrow-like bill. Thrashers spend much of their foraging effort on the ground "thrashing" through the leaf litter, gravel, and soil with their powerful beaks, in search of large spiders and insects (beetles, Jerusalem crickets, etc.). Like other mimids, the California Thrasher will consume fruit when available and is apparently particularly fond of the berries of poison oak (and aids in its dispersal), but it also takes toyon berries, elderberries, blackberries, coffeeberries, and probably any other fruit available in its habitat. The voice is what most distinguishes this species from any other and, because it often sings from an exposed perch atop a shrub, is often the first clue to its presence. The song is a clarion and "cheerful" sequence of musical phrases of varying pitch and tone, loud and continuous. Both male and female sing, and like that other chaparral songster, the Wrentit, thrashers may sing year-round.

Family Motacillidae

This is mostly an Old World family containing pipits and wagtails. Of the 65 (or so) species worldwide, only two nest in North America. Of the seven species recorded in California, only one is common, the others exceedingly rare. These are slender, long-tailed terrestrial species, adapted to foraging on open or barren ground. Their feet are rather large with an extended hind toe. All are insectivores. They are excellent fliers and many of the more northerly distributed species are highly migratory.

American Pipit ***Anthus rubescens***

LENGTH 6 IN.

Pl. 118

American Pipits are found along the coast only from September to April, walking along the ground, often at the water's edge, pumping their tails up and down. (Indeed, this bird was long known as the Water Pipit.) The habitual tail bobbing is a characteristic family trait and, along with the slender body and thin bill, a useful clue to identification. The American Pipit is a relatively plain bird, but the yellowish-buff breast overwashed with bold brown streaks is also helpful in identification. There is considerable plumage variation among the four subspecies, but the *A. r. pacificus* group, most common in our region, is the palest of the lot. Pipits tend to occur in small flocks but sometimes number into the dozens. They also frequent plowed fields and short-grass prairie and frequently consort with Horned Larks in mixed flocks. Like many small, flocking ground birds, pipits have white outer tail feathers. Why might that be?

During the breeding season, American Pipits retreat to alpine habitat above the tree line in California and other western states or northward to the Arctic tundra to nest.

Family Parulidae

This New World family of wood warblers includes over 100 species (about half of which occur in North America) of small-bodied (4 to 7.5 in.), slender-billed, insect-gleaning songbirds. Most of the warblers have lovely, melodious voices. Not only are their songs beautiful, but in breeding plumage their dress, especially that of the male, is strikingly colored and brilliant, hence the sobriquet "butterflies of the bird world." Many of the species that nest in North America migrate far south for winter, but in our temperate coastal climate, we are fortunate to have several species resident or overwintering, most reliably the Common Yellowthroat *(Geothlypis trichas)*, Yellow-rumped Warbler *(Setophaga coronata)*, and Townsend's Warbler *(Setophaga townsendi)*. Less common in winter but occasionally encountered is the Orange-crowned Warbler *(Oreothlypis celata)*, a common nesting species that typically returns in very early spring. Another common nesting species in coastal forests is Wilson's Warbler *(Cardellina pusilla)*, although it arrives about a month later than the Orange-crowned. MacGillivray's Warbler *(Geothlypis tolmiei)* is rather rare along the coast, with a patchy distribution, but is seen more frequently during fall migration.

Common Yellowthroat ***Geothlypis trichas***

LENGTH 5–6 IN.

Pl. 119

Although most of the North American warblers live in forested habitats, the yellowthroats have adapted thoroughly to freshwater and brackish marshes, swales, soggy bottomlands, and associated edge vegetation—willows, sedges, tules, cattails, gumplant. Yellowthroats tend to skulk through low vegetation, but males occasionally sing from exposed perches, especially when declaring their territory from mid-March into midsummer. But like Marsh Wrens, with whom they often share habitat, yellowthroats are more likely heard than seen. Listen for their early spring song "wich-i-ty wich-i-ty wich-i-ty" from marshes and swales. The chip note, or location call, is a distinctive "tchat," almost a spitting sound. If you catch a glimpse, the black "bandit" mask of the adult male is distinctive; both male and female have a brilliant lemon-yellow throat.

Several subspecies occur in our region: the "San Francisco" Common Yellowthroat *(G. t. sinuosa)* is limited to the Bay Area counties, where it is a resident year-round. Loss of wetland habitat has caused a severe population decline of *sinuosa* over the past century or so. Other subspecies are migratory.

SPECIAL STATUS *G. t. sinuosa* is a Bird Species of Special Concern, priority 3.

Family Emberizidae

Like the hummingbirds and warblers, this group of towhees, sparrows, juncos (pl. 123), and buntings, often called sparrows and their allies ("emberizids"), originated in the New World. However, they have radiated northward and westward to inhabit Asia, Europe, and Africa. Emberizids have relatively long tails, short wings, and stout, conical beaks adapted to eating seeds, though insects, berries, and other vegetation are also taken. Though they are sometimes referred to as New World finches, the name is misleading because they are not closely related to the finches. Although they inhabit various environments, most species are associated with shrubby places. They tend to nest low in the vegetation, or even on the ground. It is difficult to characterize their coloration; they range from dull browns to brilliant blues. Size varies from rather small (buntings) to medium (towhees, pl. 130).

Song Sparrow ***Melospiza melodia***

LENGTH 6 IN.

Pl. 120

Song Sparrows are common year-round residents of most coastal habitats, including riparian forest coastal scrub, dune grass, and freshwater and brackish marshes. These birds are not shy and often forage or sing out in the open, atop a bush, cattail, or reed, but rarely more than a few meters above the ground. Many populations are quite sedentary, that is, they remain in one area year-round. This "bottom-heavy" behavior creates discrete populations that have diverged (radiated) into numerous subspecies. In the San Francisco Bay Area, the tidal marsh Song Sparrows include three distinct subspecies, each restricted to a geographical area—Suisun Bay, San Pablo Bay, or the southern part of San Francisco Bay—and each of which differs from its neighbors in bill size, wing measurement, or other characteristics.

SPECIAL STATUS Each of the three Song Sparrow subspecies is a Bird Species of Special Concern, priority 3.

Savannah Sparrow ***Passerculus sandwichensis***

LENGTH 5 IN.

Pl. 121

This bird is much like the Song Sparrow in appearance, but paler, shorter tailed, and finer billed, usually with a wash of yellow in the face. Its buzzy, insect-like song is emblematic of coastal grasslands, short-grass prairie, and salt-marsh edges in early spring. The Savannah Sparrow is also common in coastal scrub and sand dunes, where it nests on or close to the ground in clumps or vegetation. It is often seen perched on fences. Although some Savannah Sparrows are resident year-round,

many more come from the north and east to spend the winter here. There are more than a dozen subspecies of Savannah Sparrow, varying in size and pigmentation throughout the range, especially along the coast of the western states. Separation by plumage of most subspecies is puzzling, but the resident breeding population along Northern California's coastal strip from Humboldt to Monterey County is assigned to "Bryant's" Savannah Sparrow (*P.s. alaudinus*), a California endemic.
SPECIAL STATUS *P.s. alaudinus* is a bird Species of Special Concern (year-round), priority 3.

Golden-crowned Sparrow *Zonotrichia atricapilla*

LENGTH 7 IN.

Golden-crowned Sparrows are citizens of the west coast of North America only, nesting from Alaska south through British Columbia, with most of the world's population wintering on the Pacific slope of California. Abundant at coastal California in winter, all Golden-crowneds depart for northern latitudes by the end of April, but before they leave, hormonal excitement about breeding causes males to sing. When they return in late September, after a summer absence, they may still be singing—a descending whistled "oh dear me"—revealing their presence. Wintering Golden-crowned Sparrows often gather in foraging flocks with White-crowned Sparrows, providing an opportunity for the curious naturalist to study the differences between these common *Zonotrichia*.

White-crowned Sparrow *Zonotrichia leucophrys*

LENGTH 6.5 IN.

Pl. 122

The White-crowned Sparrow is often the first land bird noticed on birding trips to the coast, seen foraging for spilled snacks in the parking lots of parks, lawns around picnic tables, and visitor's centers. Adult White-crowned Sparrows have black-and-white crown stripes; immature birds have brown-and-tan stripes. But young or old, all of them have brightly colored yellow or orange bills. Indeed, bill color is a good field mark to distinguish them from the similar Golden-crowned Sparrow in basic (winter) plumage. One subspecies, "Nuttall's" White-crowned

Sparrow *(Z. l. nuttalli)*, is a year-round resident. Another, the "Puget Sound" White-crowned Sparrow *(Z. l. pugetensis)*, nests far up the northwest coast of British Columbia and spends the winter here. The Nuttall's nests only *very* near the coast in the misty summer coolness of California's fog belt. By nesting on this edge, it replicates weather conditions of, say, Queen Charlotte Island, avoids the perils of long-distance migration, and has a nest period that is 10 weeks longer than its migratory congeners. Nuttall's White-crowned is famous for it song, with different but neighboring populations singing a variety of dialects, each distinguishable from the other, even to human ears. Most distinctive and most variable among populations are the end notes following the pure tones of the clarion introductory notes.

Family Cardinalidae

This diverse group (11 genera, 42 species) sports some of the most striking plumages within the North American avifauna. The males are often brilliantly colored with blazing reds, lemony yellows, and cerulean blues. They are mostly seed-eating, forest-dwelling species with powerful, sometimes massive, beaks. The diversity of the family in our region is exemplified by three of the most common representatives: Black-headed Grosbeak *(Pheucticus melanocephalus)*, Western Tanager *(Piranga ludoviciana)*, and Lazuli Bunting *(Passerina amoena)*. The rather large grosbeak with its outsized bill and dazzling plumage is a seed-eating bird that occupies an eclectic choice of habitats—riparian corridors, conifer forest, deciduous woodland, oak savannah, and residential developments. Its primary requirement seems to be a well-developed understory with a treelike overstory. The tanager is more refined in its choice of forest—open (but not too open!) conifer and mixed coniferous and deciduous woodlands—and eats mostly insects but takes fruits and berries when available. The bunting is a bird of shrublands, brushy grasslands, or riparian edges where it forages close to the ground, gleaning insects from foliage and grasses, picking berries from the bush, or even "flycatching" from a low perch. Grosbeaks are the most common of the three in our region; tanagers and buntings can be patchily distributed through coastal habitats but may be found at migration sites, especially in fall.

Family Icteridae

Many taxonomists consider this New World family of blackbirds and orioles the "newest," or most recently evolved, family of birds. Its 97 species in 27 genera show a wide range of behavioral and morphological variation. Many species have some prismatic, reflective black in the plumage. Family members may be migratory or sedentary, solitary or colonial, drab or spectacular. Sexual dimorphism in both plumage and size is common. In a study of the family, Lowther (1975) found that plumage dimorphism is more common at higher latitudes; size dimorphism more commonly occurs at lower latitudes.

Western Meadowlark ***Sturnella neglecta***

LENGTH 9.5 IN.

Pl. 124

The name is something of a misnomer, for this colorful songster of grasslands and open country is not a lark, but a cousin to the blackbirds and orioles—an "icterid." The black V-shaped necklace (pectoral crescent) set off by the brilliant lemon-yellow throat and breast is distinctive. Sexes are alike, though males may be brighter. The body build is rather sturdy, the tail rather short, and the bill slender and pointed. This is a grassland specialist, a ground forager of open country—pastures, prairies, grasslands, and barrier dunes. Like some other flocking birds of open country, Western Meadowlarks flash their white outer tail feathers when they fly.

Western Meadowlarks are sparsely strewn through coastal habitats in the nesting season, but thereafter, they gather in small flocks and become rather common on the coastal prairie from September through March. The Eastern Meadowlark does not occur in California, therefore identification of the Western Meadowlark is easy.

This bird is not only handsome, but also melodious. The complex warbling song of the adult, whether given in flight or from a perch on a fencepost, earned "Meadowlark-man" a reputation as an unrepentant gossip in Native Californian legend, known for often making uncomplimentary remarks.

Brown-headed Cowbird ***Molothrus ater***

LENGTH 7 IN.

Pl. 125

Bare-ground specialists, Brown-headed Cowbirds evolved on the Great Plains of North America but followed European expansion westward in the 19th century and colonized California through the 20th century, arriving on the northern coast in the 1940s. This turn of events, concurrent with large-scale modification of the landscape, especially through agricultural and forestry practices, has been disastrous for native songbirds. Cowbirds are brood parasites, laying their eggs in the nests of host species, which feed and raise the cowbird nestlings. Riparian songbirds have been particularly hard hit by this invasion, but so have a myriad of other coastal birds (e.g., White-crowned Sparrow). Cowbirds often consort with blackbird flocks on plowed fields and barren ground in the nonbreeding season, or they gather in feed lots where they are subsidized through the winter, then venture out in search of foster parents (hosts) for their eggs in the spring.

Red-winged Blackbird ***Agelaius phoeniceus***

LENGTH 7–8 IN.

Pl. 126

Gregarious and abundant at farms and ranches on the coast, Red-winged Blackbirds use freshwater or brackish marshes for nesting and communal roosts, but may also nest in upland weed patches, hayfields, or drainage canals. Like Marsh Wrens, with which they share nesting grounds, Red-wingeds are polygamous—the male mates with several females that nest within a territory (a patch of marsh) that he defends against other males. This male does not know (or care) where the various nests are and never helps with domestic chores (see Tricolored Blackbird). Evolution seems to have selected male Red-wings, as well as their neighboring male Marsh Wrens, as the exemplary male chauvinists.

The singing male Red-wing, issuing a guttural, echoic "oak-a-lee" from atop a cattail or bulrush while inflating his brilliant red epaulets, is as evocative of freshwater wetlands as any image imaginable. Interestingly, males in coastal California (and the

Central Valley) lack the yellow border of epaulet familiar in most other subspecies and are known locally as the "bicolored redwing." The absence of a yellow border to the epaulet may be an adaptation to species identification, caused by overlap with the Tricolored Blackbird (Orians and Collier 1963), which has white bordering the epaulets.

Tricolored Blackbird ***Agelaius tricolor***

LENGTH 7–8 IN.

Pl. 127

Tricoloreds are virtually endemic to California, and the world population is probably less than 150,000 individuals, exhibiting a declining trend throughout the state. Tricoloreds are not as ubiquitous as Red-winged Blackbirds, but they are highly sociable, so when found, they are usually in large groups.

Tricolored Blackbirds are colonial and less promiscuous than their Red-winged cousins. Although they may be polygynous, many are monogamous and males help the females in care and feeding of the young. Tricolors are extremely gregarious, nesting in large colonies; in fact, they form "the largest colonies of any North American passerine" (Beedy and Hamilton 1999). But they are itinerant nesters, nesting at more than one locale in a given season and moving to different nesting locations between years, an unusual adaptation to ephemeral food sources.

Winter flocks of Tricoloreds are found predictably in pasturelands, often foraging on barren ground, from Castroville to Point Reyes (where 8,000 to 12,000 have been found annually) and are generally rare north of the Russian River. Single, small nesting colonies have been active, however, at Fort Bragg and at Humboldt Bay near Fortuna.

SPECIAL STATUS Bird Species of Special Concern, priority 1.

Brewer's Blackbird ***Euphagus cyanocephalus***

LENGTH 8–9 IN.

Pl. 128

Brewer's Blackbirds are common year-round residents in our region. There are always some among the large flocks of Red-winged and Tricolored Blackbirds, and some flocks are composed entirely of Brewer's.

Scattered individuals are often seen marching around stables, picnic tables, parking lots, and even sandy beaches, where they forage for snacks and scraps, spilled or offered by humans. The iridescent shining plumage and glaring yellow irises of the male are unmistakable. The more cryptic female could be mistaken for a Brown-headed Cowbird. The breeding display of the male—strutting and puffing its iridescent feathering—is a common sight in city parks, a bit of natural beauty in the human-altered landscape.

Family Fringillidae

Finches are smallish, gregarious, seed-eating passerines with stubby bills. Their plumage is highly variable, and most species are dimorphic. Sixteen species have been recorded in California, and five occur regularly in our region. Two other species are irruptive, present some years and hard to find or virtually absent in other years: the Red Crossbill *(Loxia curvirostra)* and Evening Grosbeak *(Coccothraustes vespertinus).*

The Purple Finch *(Haemorhous purpureus)* (pl. 131) and the House Finch *(H. mexicanus)* are similar in appearance but differ in habitat associations, for the most part. The Purple Finch is a bird of the coniferous forest, though it occupies a wide range of habitats during the nonbreeding season. The House Finch, originally a desert species, has invaded just about every habitat in North America over the last century, except for dense conifer forests. It is essentially an edge species, and particularly common in human-altered, urban and suburban habitats.

Three of our most common fringillids are in the same genus *(Spinus).* They are all small, seed eating, and gregarious and are most often found in flocks, even during the nesting season. All have undulating flight and bright wing patches that flash in flight. The Pine Siskin *(S. pinus)* has the slenderest bill of the group, an adaptation to extracting seeds from grasses, thistles, and alder cones. It is a gregarious species, often seen in large flocks, hanging from branches and chattering as they feed, or bouncing through the air at canopy height. Siskins are nomadic, wandering around looking for seed crops, and irruptive, with populations responding to the abundance of seeds available.

The Lesser Goldfinch *(S. psaltria)* (pl. 132) is not really migratory, but some seasonal movement occurs when the birds tend to withdraw from the Coast Ranges during the coldest and most inclement weather.

The American Goldfinch *(S. tristis)* is more widespread and more familiar than its slightly smaller cousin. This "wild canary" is an abundant nesting species all along the coast and is easily found from early April into October. Like the Lesser Goldfinch, it moves south and inland before winter. Spring males are bright yellow and flamboyant (rather than furtive, like warblers) with unmistakable plumage, although the duller winter plumage of both male and female may confuse the novice observer. Bill shape is a helpful field mark—conical and stubby, but sharp—the perfect tool for extracting seeds from thistle flowers. Small groups are often seen feeding at thistles or in coastal scrub. These birds are quite vocal—especially during their undulating flight. The three- or four-note flight call (contact call) is emphatic and sounds like "tato-chip" or "potato-chip"—easy to remember because they fly with a dip.

BIRDING OPPORTUNITIES AND ROADSIDE NATURE CENTERS

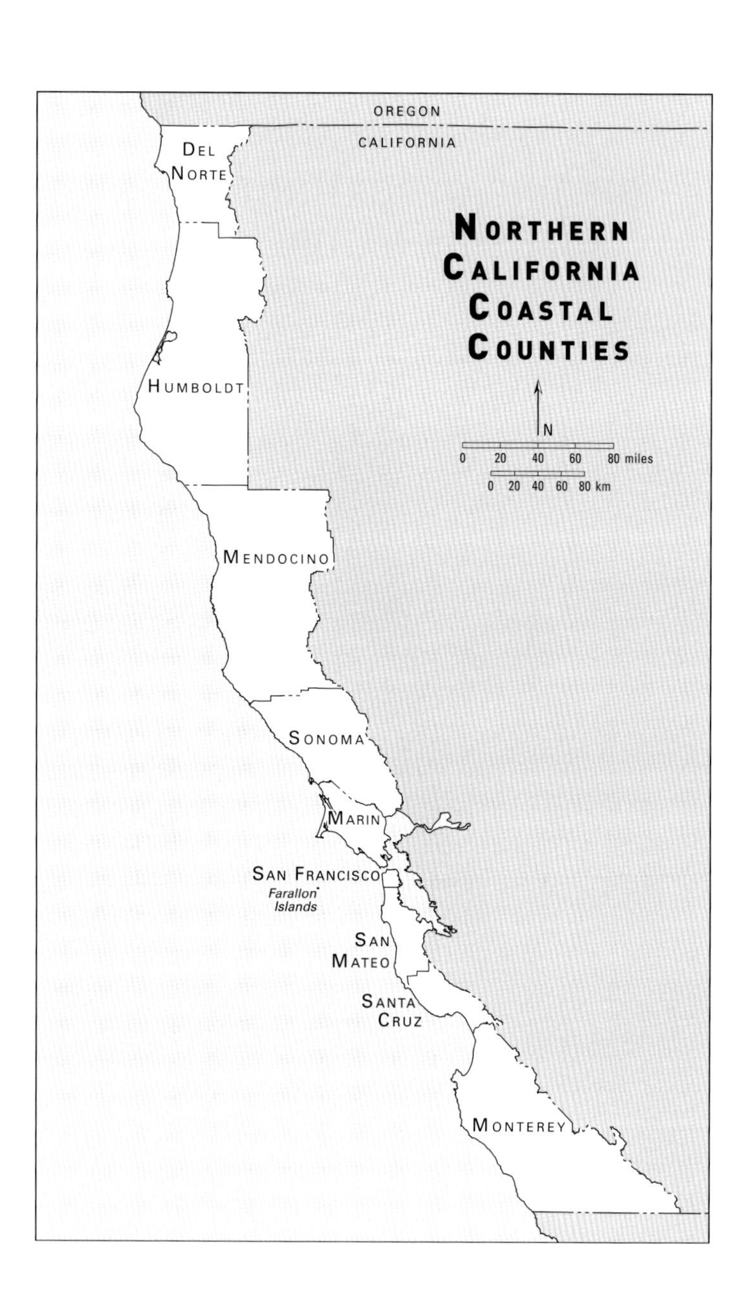
OREGON
CALIFORNIA
NORTHERN CALIFORNIA COASTAL COUNTIES
N
0 20 40 60 80 miles
0 20 40 60 80 km
DEL NORTE
HUMBOLDT
MENDOCINO
SONOMA
MARIN
SAN FRANCISCO
Farallon Islands
SAN MATEO
SANTA CRUZ
MONTEREY

Contact information is given for each location listed here, when available, but it is advisable to check before your visit, because hours of operation, as well as phone numbers and websites, are always subject to change.

This guide provides an overview of some of the most important and productive birding spots along the coast, from the south to the north, but there are numerous other locales to explore. Local knowledge is always your best source of information, and most local Audubon Societies or regional bird clubs post site guides and current information about bird occurrence and distribution for their specific areas on their websites. When exploring a region, it is always advisable to check the websites and local bird guides for more in-depth guidance. Most locales have Internet listserves that post unusual sightings or phenomena.

Monterey County

The Big Sur Ornithology Lab (BSOL)

VENTANA WILDERNESS SOCIETY, BSOL, COAST ROUTE, HC67 BOX 99, MONTEREY, CA 93940

(831) 624-1202; INFO@VENTANAWS.ORG

WWW.VENTANAWS.ORG/CONSERVATION_BSOL/BIRD_BANDING_LAB.HTM

OPEN APRIL–AUGUST THURSDAYS AND SUNDAYS. HEAVY RAIN CANCELS ACTIVITIES.

AMENITIES INCLUDE RESTROOMS, CAMPING, HANDICAPPED ACCESS.

THERE IS A FEE TO ENTER THE STATE PARK.

The BSOL is located in Andrew Molera State Park, 22 mi south of Carmel Crossroads (Rio Road and Hwy 1) on Hwy 1. Just south of the Point Sur Lighthouse, take the entrance to Andrew Molera Park on the west side of the highway. Go 50 yd downhill and turn left just before the parking lot entrance. Go about 100 yd down this dirt road to the BSOL parking lot on the left.

Staff and interns at the Banding Lab capture, examine, and band wild songbirds to gather scientific and life history information. Visitors and small groups are welcome anytime during banding operations for the first five daylight hours, beginning 15 minutes after sunrise. Groups of more than six people should make prior arrangements.

The banding routine and data recording is of great interest

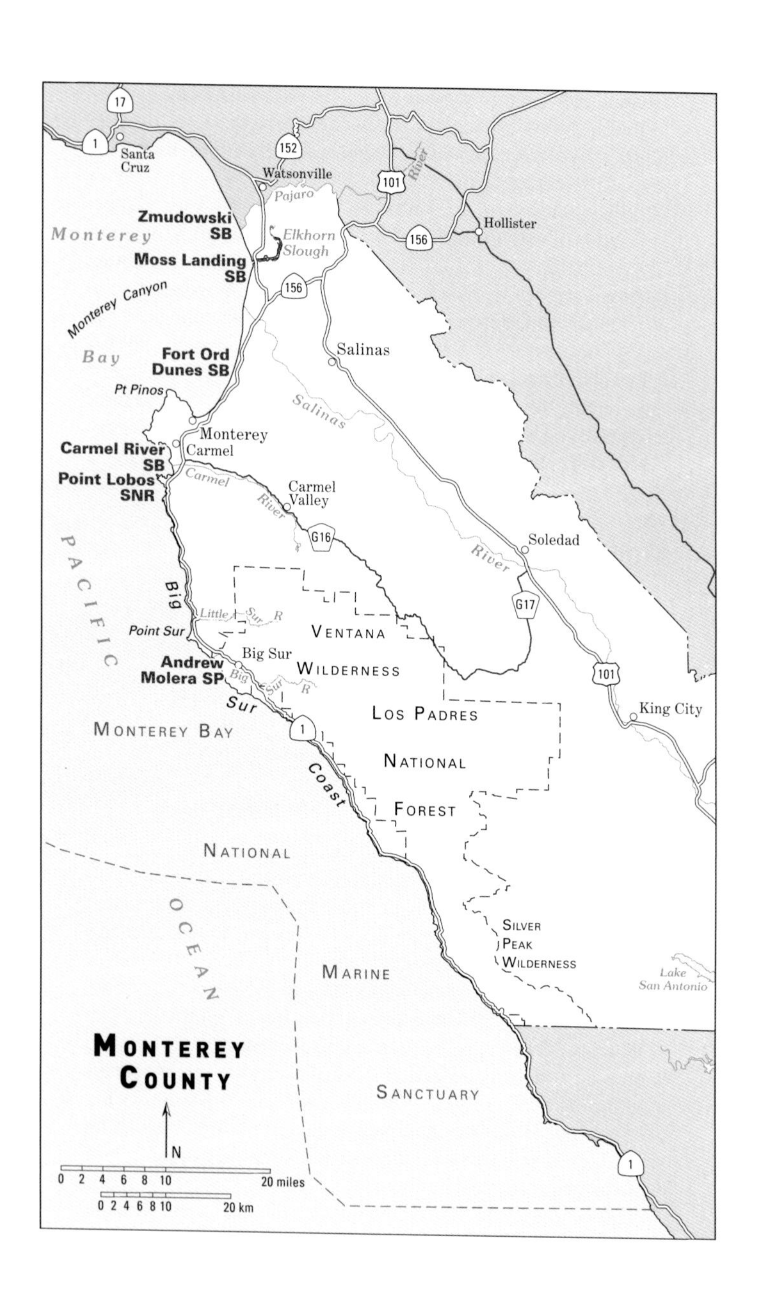
17
1
Santa Cruz
152
Watsonville
101
River
Pajaro
Hollister
156
Zmudowski SB
Monterey
Elkhorn Slough
Moss Landing SB
156
Monterey Canyon
Salinas
Bay
Fort Ord Dunes SB
Pt Pinos
Salinas
Monterey
Carmel River SB
Carmel
Point Lobos SNR
Carmel
Carmel Valley
River
G16
Soledad
PACIFIC
River
Big
Little Sur R
G17
Point Sur
VENTANA
Big Sur
Andrew Molera SP
WILDERNESS
Big Sur R
101
Sur
King City
LOS PADRES
MONTEREY BAY
1
NATIONAL
Coast
FOREST
NATIONAL
OCEAN
SILVER PEAK WILDERNESS
MARINE
Lake San Antonio
MONTEREY COUNTY
SANCTUARY
N
0 2 4 6 8 10 20 miles
0 2 4 6 8 10 20 km
1

to anyone who is interested in wild birds, and the banders are anxious to share information. Having a wild bird held up to one's ear is an unforgettable experience—the sound or "feeling" of wild energy running through a bird is an experience not to be missed.

The greatest bird diversity is from September through April, but there is plenty of activity for the curious naturalist at any time of year. Several trails originate at the lab to meander through oak and riparian woodlands, past sycamores, to a lagoon, a barrier beach, and headlands with ocean views. A seasonal footbridge crosses the Big Sur River. The beach and the shallow lagoon near the river mouth attract a variety of shorebirds and gulls. Headlands Trail leads from the beach up to Molera Point and provides a vantage point from which to scope for whales and marine birds. Most of the land-birding here (except birds in the hand) is in varyingly dense vegetation. Seeing many of the insectivores (leaf gleaners) and ground dwellers (seed eaters) can be difficult and requires some patience.

The Monterey Bay Aquarium

886 CANNERY ROW, MONTEREY, CA 93940

(831) 648-4880; (800) 840-4880; (800) 648-4860 TICKETS, SPECIAL RATES, AND GUIDED TOURS; (831) 648-4888 GENERAL INFO

WWW.MONTEREYBAYAQUARIUM.ORG

AMENITIES INCLUDE RESTROOMS, FOOD, STORES, HANDICAPPED ACCESS.

THERE IS A SIGNIFICANT ENTRY FEE.

RESERVATIONS MAY BE NECESSARY, SO CALL AHEAD.

"THE MISSION OF THE MONTEREY BAY AQUARIUM IS TO INSPIRE CONSERVATION OF THE OCEANS."

Access is from Hwy 1 (110 mi south of San Francisco) to Monterey Harbor; then follow the signs to Cannery Row or the aquarium, very popular destinations.

The Monterey Bay Aquarium is a world-renowned facility in a stunningly beautiful setting, featuring mostly local wildlife from the nearshore Pacific. Nearly 200 living exhibits, hands-on displays, and spectacular dioramas provide marvelous learning experiences for everyone. There are always expansive displays with rotating themes such as "jellies" or sharks. The large aquarium requires at least a half day to experience fully. Families and well-behaved groups of kids will have an especially fine time. While there is not an emphasis on birds, there are live sandpipers and plovers in an open-air aviary, and the other exhibits are closely related to birds and ocean habitats.

Bring your binoculars to the aquarium, as some great bay birds can be seen right off the back porch. Look for Brandt's and Pelagic Cormorants, Peregrine Falcon, Black Oystercatcher, and Black Turnstone, loons, grebes, Elegant Terns (summer), Common Murres, and Pigeon Guillemots.

Back outside, take Ocean View Avenue west from the aquarium along the Pacific Grove shore, around Point Pinos and to Asilomar Beach. There are many turnouts where you can scan Monterey Bay, intertidal rocks, and sea stacks. The scenery is always captivating, and wild surf is usually spectacular.

Elkhorn Slough National Estuarine Research Reserve

ESNERR VISITOR CENTER, 1700 ELKHORN ROAD, WATSONVILLE, CA 95076

(831) 728-2822; INFO@ELKHORNSLOUGH.ORG

WWW.ELKHORNSLOUGH.ORG

VISITOR CENTER AND RESERVE GROUNDS ARE OPEN WEDNESDAY–SUNDAY 9–5.

ADMISSION TO THE VISITOR CENTER AND PICNIC TABLES IS FREE, BUT THERE IS A $2.50 DAY-USE FEE FOR TRAIL HIKING (16 YEARS AND OLDER; CHILDREN ARE FREE), WHICH IS A REAL BARGAIN FOR THEIR MILES OF TRAILS.

GREAT VISITOR CENTER, BATHROOMS, PICNIC TABLES, NATURALISTS TO HELP PLAN YOUR VISIT OR TO ANSWER YOUR QUESTIONS, AND DOCENT TOURS, AND IN CASE YOU FORGOT YOUR BINOCULARS OR BIRD BOOK, THEY ARE AVAILABLE FOR LENDING.

The visitor center is located at the east end of Elkhorn Slough Preserve, about 5 mi from Moss Landing on Hwy 1. Coming from north or south on Hwy 1, turn east on Dolan Road at Moss Landing at the giant smokestacks. Go 3.5 mi on Dolan, turn left onto Elkhorn Road. Go 2.2 mi on Elkhorn and turn left into the ESNERR Visitor Center.

The newly renovated visitor center has award-winning exhibits, mostly considering the ecology of life and tide flow in estuaries and the marine environment. Nature walks are led by volunteer docents every Saturday and Sunday at 10 AM and again at 1 PM, and there is an early bird special walk that departs at 8:30 AM the first Saturday of every month. Much can be learned at the visitor center, maybe more out on the trails.

This is a favorite destination when birding the Monterey Bay environs. Elkhorn Slough itself, and the adjacent oak-savannah habitats accessed by trails from the visitor center, afford excellent opportunities to find birds. Additionally, the harbor at Moss Landing, especially along Jetty Road (about 0.5 mi north of the Elkhorn Slough Bridge on Hwy 1) is bird wealthy any time of year. The harbor is tidal (as is the slough) and from late June through mid-April is a great place to see many shorebird species, close up. May and June are the times of lowest bird activity, but even then some stunning waders are present, especially American Avocets and Black-necked Stilts that nest nearby. Great Blue Herons and Snowy and Great Egrets are present throughout the year.

In late summer, hundreds or thousands of Brown Pelicans, Heermann's Gulls, and Elegant Terns forage offshore and roost on harbor sandbars or across the highway on levees in the Moss Landing Wildlife Area. Winter birds include loons (three species), grebes, cormorants, scoters, Red-breasted Mergansers, and other sea ducks, many gulls of several kinds, and Forster's Terns.

Santa Cruz County

A thorough guide to Santa Cruz is available online through the Santa Cruz Bird Club website. The coastal county environs are partitioned into three regions: (1) south county (Pajaro Valley), (2) mid county, and (3) north county. Select sites from each area are given here.

South County: Pajaro Dunes and Pajaro River Mouth

Take the Riverside Drive (Hwy 129) exit west off Hwy 1 to the intersection with West Beach Road. Turn left (west) on West Beach Road and drive 2.8 mi to its end, and park. The mouth of the Pajaro River is a 1 mi walk south along the beach.

This is arguably the best place in the county for waterbirds, and all of the shorebirds one might expect on a sandy beach occur hear, as well as a generous variety of gulls (including Thayer's Gull in winter), terns, and diving birds. In late summer and early fall, when Elegant Terns are moving up the

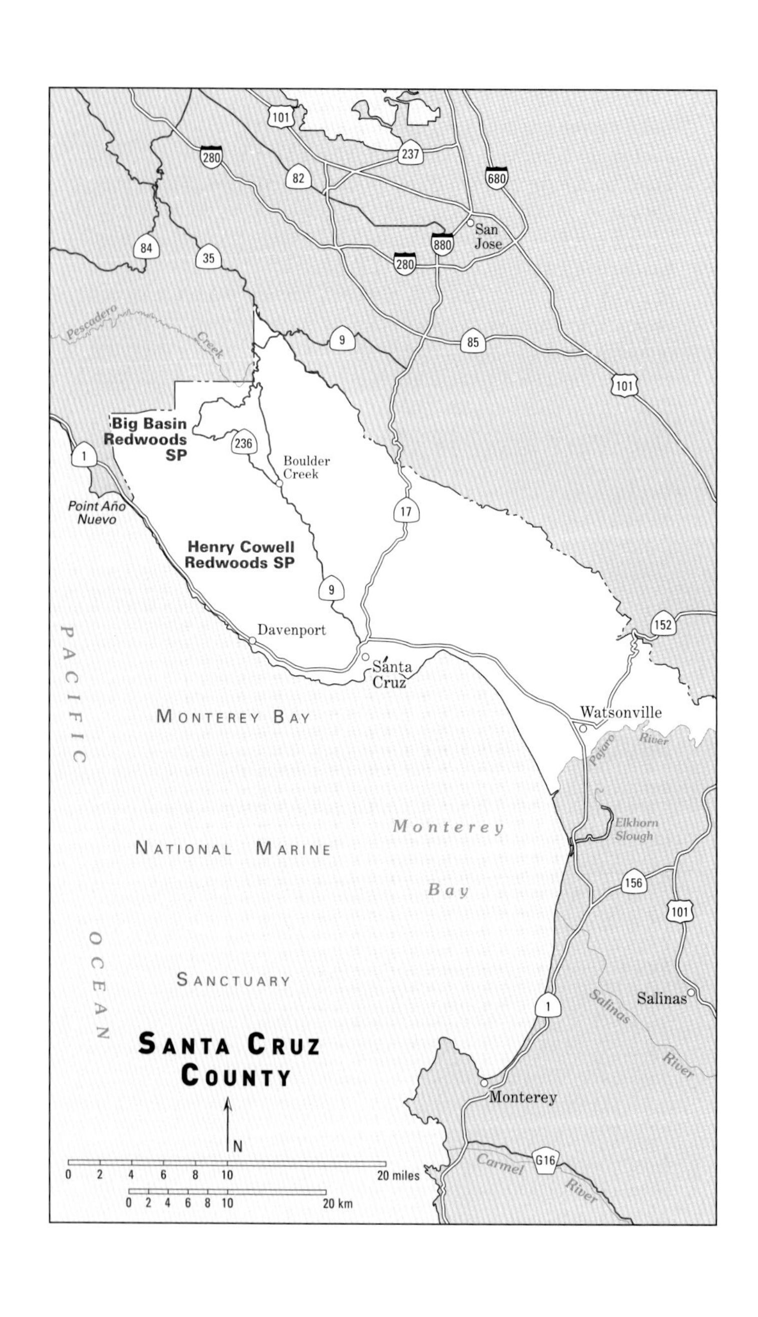

101
280
237
82
680
San Jose
880
84
35
280
Pescadero Creek
9
85
101
Big Basin Redwoods SP
236
1
Boulder Creek
Point Año Nuevo
17
Henry Cowell Redwoods SP
9
Davenport
152
Santa Cruz
PACIFIC
MONTEREY BAY
Watsonville
Pajaro River
Elkhorn Slough
Monterey
NATIONAL MARINE
156
Bay
101
OCEAN
SANCTUARY
1
Salinas
Salinas River
SANTA CRUZ COUNTY
Monterey
N
0 2 4 6 8 10 20 miles
G16
Carmel River
0 2 4 6 8 10 20 km

coast, Parasitic and Pomerine Jaegers are often shadowing them, opportunistic pirates that they are. This is an important nesting site for the threatened Western Snowy Plover, so please abide by the signs designating closure areas during the nesting season.

After returning to your car at West Beach Road, head back eastward 0.2 mi and turn left on Shell Road. In a short way you will cross Watsonville Slough. Park in the pullout on the right shortly beyond the slough crossing outside the entrance to the Shorebirds development. Check the slough for waterfowl and waders, then take the path around the north side of Shorebird Pond to the southern end of Sunset State Beach.

Mid-County: Neary Lagoon

Riparian thickets and freshwater marsh habitat have been encroached upon in Santa Cruz County as elsewhere, therefore remnant patches of these habitats become even more important. Historically, Neary Lagoon has been one of the best freshwater sites in the county, and although development has intruded, it still supports a complement of wetland birds.

An access point off California Street, east of Bay Street, in Santa Cruz town, leads to paths and wooden walkways through the marsh, allowing an intimate walk within the wetland vegetation. This is a good spot to check for land birds during migration, and the lagoon is visited by herons and egrets, freshwater ducks and rails, and of course those three common marsh denizens—Marsh Wrens, Common Yellowthroat, and Song Sparrow.

North County: Waddell Creek Beach at Big Basin Redwoods State Park

The ocean edge of Big Basin Redwoods State Park has a protected sandy beach, a brackish marsh, and some riparian thickets. Parking is available along the side of the highway in the vicinity of the Waddell Creek bridge. The Rancho del Oso History and Nature Center, open only on weekends, has a marked driveway off Hwy 1 just south of the Waddell Creek bridge. The lagoon at the mouth of the creek attracts shorebirds, as does the nearby beach. This is a favorite destination for windsurfers and surfers, but also a fairly reliable place to see Marbled Murrelets just offshore, year-round. These charming and threatened little seabirds nest upstream in the redwood forests and can be heard, and even seen, as they fly over, visiting the canyon at dawn in summer. On the east side of the highway, Marsh Trail begins at the Nature Center and crosses the marsh to the Rancho del Oso office. The marsh is worth checking for Wood Ducks, mergansers, and other freshwater marsh species.

San Mateo County

Año Nuevo State Reserve

NEW YEAR'S CREEK ROAD, PESCADERO, CA 94060
(650) 879-0227 TAPED INFORMATION, (650) 879-2025 ENTRANCE KIOSK
WWW.PARK.CA.GOV

The reserve is located just west of Hwy 1, 5 mi south of Pigeon Point and just north of the San Mateo–Santa Cruz County line. Take Hwy 1 about 25 mi south from Hwy 92 at Half Moon Bay or 16 mi north from Santa Cruz. The entrance, on the west side of Hwy 1, is well signed. The reserve is open for day use only, with a small fee; no camping, dogs, or bicycles are allowed. There is a visitor's center with informative exhibits and docent-led tours to the seal haul-out from November to April. During the seal season, access to the point is allowed *only* by docent-led tours.

What is here? While the Northern Elephant Seal *(Mirounga angustirostris)* is the major wildlife attraction, alert birders record many interesting species, any time of year, during a hike from the parking lot to the point. The beautiful and endangered San Francisco Garter Snake *(Thamnophis sirtalis tetrataenia)* is here in small numbers and may be seen (by fortunate visitors) basking on or beside the trail on sunny days. Elephant Seal docents will tell you all about the lives of these huge pinnipeds, but you will probably have to sort out the birds on your own. A prehike stop at the visitor's center will be enlightening.

Raptors such as Northern Harrier, White-tailed Kite, and Red-shouldered and Red-tailed Hawks are often seen perched on, or flying over, coastal scrub habitats adjacent to the trails. Brandt's and Pelagic Cormorants and Black Oystercatcher are present all year and may be seen from the headlands, flying

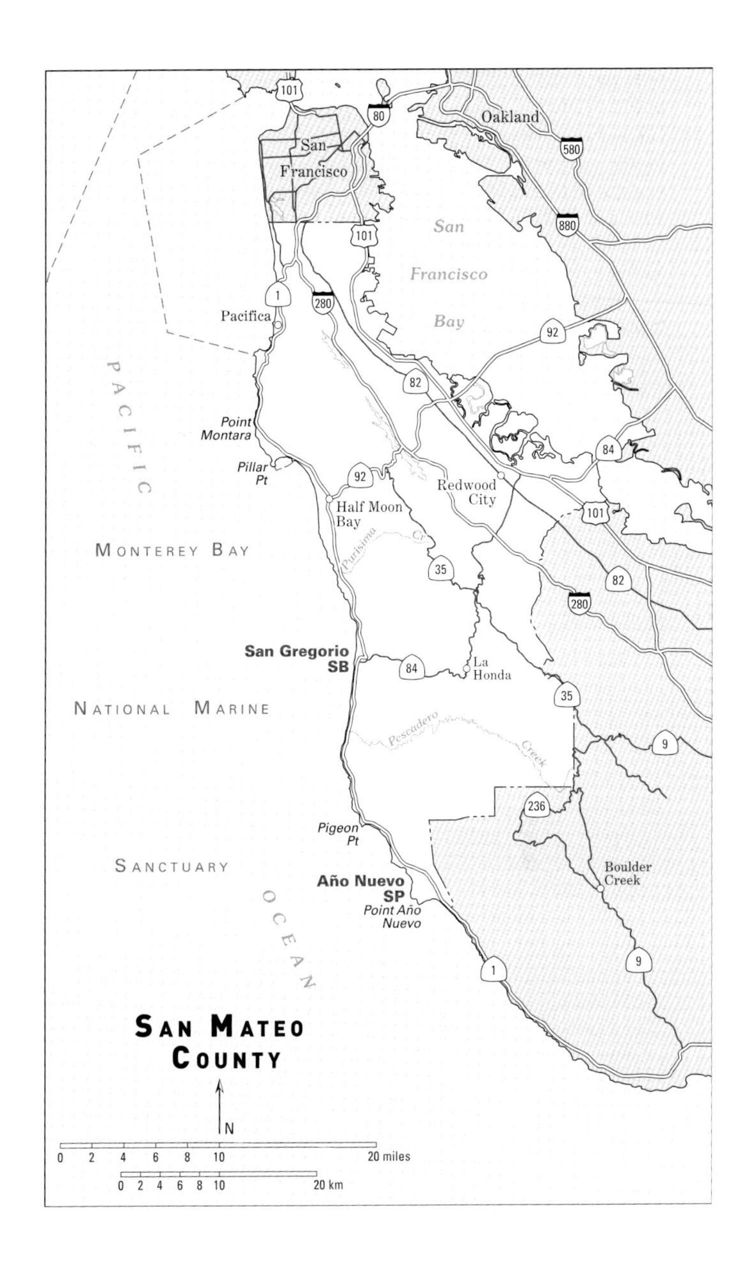

San Francisco
Oakland
San Francisco Bay
Pacifica
Point Montara
Pillar Pt
Half Moon Bay
Redwood City
Purisima Cr
La Honda
Pescadero Creek
San Gregorio SB
Pigeon Pt
Año Nuevo SP
Point Año Nuevo
Boulder Creek
PACIFIC OCEAN
MONTEREY BAY NATIONAL MARINE SANCTUARY
SAN MATEO COUNTY
101
80
580
880
1
280
92
82
84
35
9
236
N
0 2 4 6 8 10 20 miles
0 2 4 6 8 10 20 km

over the water or perched on islets or sea stacks. In late summer and fall, Brown Pelicans and Heermann's Gulls may be plentiful, perched on the headlands, while Pigeon Guillemots and grebes may be scoped on the water. The rare Marbled Murrelet is occasionally spotted here because the southernmost nesting sites are in the nearby Santa Cruz Mountains. Near the point, Sanderlings and Black Turnstones forage on sandy beaches or in the kelp piles.

There are some very special land birds present. A colony of Bank Swallows nests in holes in the sandstone cliffs, and adults may be seen flying from April into August—especially near the pond along the southernmost trail. A few pairs of California Thrashers live year-round in the willow thickets and coastal scrub. This is the only place in Northern California where this species occurs within singing distance of the sea. Another rare and elusive species, the Black Swift, nests (or did formerly) on the walls of sea caves and can (or could) occasionally be seen foraging on flying insects from May into August.

Fitzgerald Marine Reserve of the San Mateo County Parks

FRIENDS OF FITZGERALD MARINE RESERVE, P.O. BOX 451, MOSS BEACH, CA 94038
(650) 704-5809 JUNIOR RANGER PROGRAM
WWW.FITZGERALDRESERVE.ORG

This reserve is located just west of Hwy 1 at Moss Beach, a small town about 10 mi south of Pacifica and 7 mi north of Half Moon Bay. The reserve visitor's center is west a couple of blocks through residential neighborhoods, at the end of California Avenue. There are bathrooms and picnicking possibilities and a naturalist staff to talk with you about tide pool life—and, perhaps, birds.

The reserve was created to highlight an intact and diverse intertidal habitat and to protect it from "scientific" collecting and food gathering by humans. As a result, it contains some of the best tide pools in the Bay Area—a wonderful and easily accessible place to discover and observe intertidal plants and animals. There is a paved parking lot (38 vehicles), with toilet facilities and picnic tables. There is *no* entrance fee. A 1.2 mi loop trail along bluff and beach is an easy stroll. Reserve naturalists may be around to answer questions.

Except for a few oystercatchers, Surfbirds, Willets, Black Turnstones, gulls, and occasional migrant land birds around the parking lot and neighborhoods, this is not a very bird-rich destination, but it is worth a stop for anyone interested in the nature of the California coast. Harbor seals and even whales may be visible from the bluff.

NEARBY BIRDING Pillar Point Harbor is about 2 mi south of Moss Beach at the town of El Granada. South on Hwy 1, and past a small airport on the west side of the road, turn right (west) at the stoplight, then quickly left (south) into the large parking lot. During winter there are usually several species of shorebirds, grebes, and diving ducks, Common Loons, and a kingfisher. More kinds of shorebirds are possible during migration, and cormorants and Brown Pelicans roost by the hundreds on the rocky breakwaters. It's always worth scanning the breakwater for rocky shore birds. Just park and walk around. There are restaurants, public toilets with running water, and trinket shops.

Coyote Point Museum (now CuriOdyssey)

1651 COYOTE POINT DRIVE, SAN MATEO, CA 94409

(650) 342-7755; INFO@COYOTEPTMUSEUM.ORG

WWW.COYOTEPTMUSEUM.ORG

OPEN TUESDAY–SATURDAY 10–5, SUNDAY 12–5, CLOSED MONDAY

ADULTS, $6; SENIORS AND KIDS 13–17, $4; KIDS 3–12, $2. PARK ENTRY IS $4 PER CAR.

AMENITIES INCLUDE EXHIBITS, PROGRAMS, PICNIC AND GROUP PICNIC AREAS (GROUPS RESERVE AT 650-363-4012), AND BATHROOMS. THE MUSEUM IS WHEELCHAIR ACCESSIBLE, AND HANDICAPPED PARKING IS AVAILABLE ADJACENT TO THE BUILDING.

From the north, take the Poplar Avenue exit from Hwy 101. Turn right on Humboldt to Peninsula Avenue, then right again, over the freeway and circle into the park. From the south, take the Dore Avenue exit from Hwy 101. Immediately turn left onto North Bayshore Blvd, then right into the park.

The San Mateo County Park, where the museum is situated, provides easy public access to the shore of San Francisco Bay near the otherwise bustling urbanized habitat surrounding the San Francisco airport. Birding is best in winter. Plantings on the museum grounds shelter many native land birds, many of which become rather unafraid and easy to see. This oasis of greenery also attracts weary migrant land birds, sometimes including local rarities. The bay shore is limited in extent, but it does attract an array of waders, shorebirds, terns, and gulls. Black Oystercatcher and Red Knot are good possibilities here.

Alameda and Santa Clara Counties

Don Edwards San Francisco Bay National Wildlife Refuge (founded 1974)

P.O. BOX 524, NEWARK, CA 94560

(510) 792-0222; CARMEN_LEONG-MINCH@FWS.GOV

HTTP://DESFBAY.FWS.GOV

To get to the refuge headquarters and visitor's center, from Hwy 84 (at the east end of the Dumbarton Bridge), take the Thornton Avenue exit, and go south for 0.8 mi to the refuge entrance on the right. Turn in and follow Marshlands Road to the stop sign, then turn left into the parking lot. To access the refuge environmental education center from Interstate 880 (or Hwy 101), exit on Hwy 237 toward Alviso. Turn north onto Zanker Road and continue 2.1 mi to the center's entrance road (a sharp right turn at Grand Blvd).

Located on the southeast corner of San Francisco Bay near the towns of Newark and Alviso, this 30,000-acre National Wildlife Refuge protects extensive bay wetlands and adjacent uplands. Hiking is generally level and mostly handicapped accessible. Bring your own food and fluids. There are many exciting instructor-led programs and frequent field trips on topics including astronomy, birding, bats, botany, family specials, tracking, tiny wiggly-squiggly creatures, and watersheds. Great informative exhibits and friendly rangers at the visitor's center will enhance your overall experience, and there is a free Junior Naturalist summer day camp.

The refuge is a birding hot spot, attracting over 280 species of birds and protecting nesting habitat for several threatened and endangered species, most notably California Least Tern, "California" Clapper Rail, and Western Snowy Plover. Both the visitor's center and the environmental education center (different buildings at different places in the large refuge complex) house informative exhibits. Many trails and naturalist-led walks begin from the headquarters. The quarterly newsletter, *Tideline* (available online), announces numerous programs and activities.

The extensive system of trails passes by tidal marsh and salt ponds where many species of shorebirds (including American Avocets, Black-necked Stilts, and Snowy Plovers) may be closely observed; common land birds can be found in the trees and shrubs outside the visitor's center. Just to the north, Coyote Hills Regional Park encompasses a greater diversity of habitats, and more species of birds (as well as flora, reptiles, amphibians, and mammals) may be seen in a short time period.

San Francisco County

Golden Gate Park and Golden Gate National Recreation Area (Presidio and Crissy Field)

Although the city of San Francisco is, for the most part, a gritty urban landscape, it's position between a large estuary and the productive ocean afford better birding opportunities than can be found in most large cities. Most notably, Golden Gate Park, at 1,017 acres, and the Presidio lands, as well as several other smaller parks, are well worth exploring.

Golden Gate Park is on the western edge of the city, a landscaped park built on former coastal dunes. Although entirely man-made, the park now supports diverse vegetative communities and planted copses of pine, cypress, and eucalyptus. Nearly 300 species of wild birds have been recorded in the park, and it is an island of habitat amid miles of pavement, for both nesting and migrant birds. The eastern half of the park is the least birdy, but a few spots are worth visiting, the Rhododendron Dell and the Botanical Gardens (formerly known as Strybing Arboretum) in particular. The best birding opportunities are in the western half of the park, from Crossover Drive west to the Great Hwy and Ocean Beach. The various lakes in this area attract impressive numbers of waterfowl in winter, as well as egrets and herons. Chain of Lakes, between 41st Avenue and Lincoln Way, supports three small lakes that are favored by birders. Other lakes are also worth checking: Elk Glen Lake, Mallard Lake, Stowe Lake, and South, Middle, and

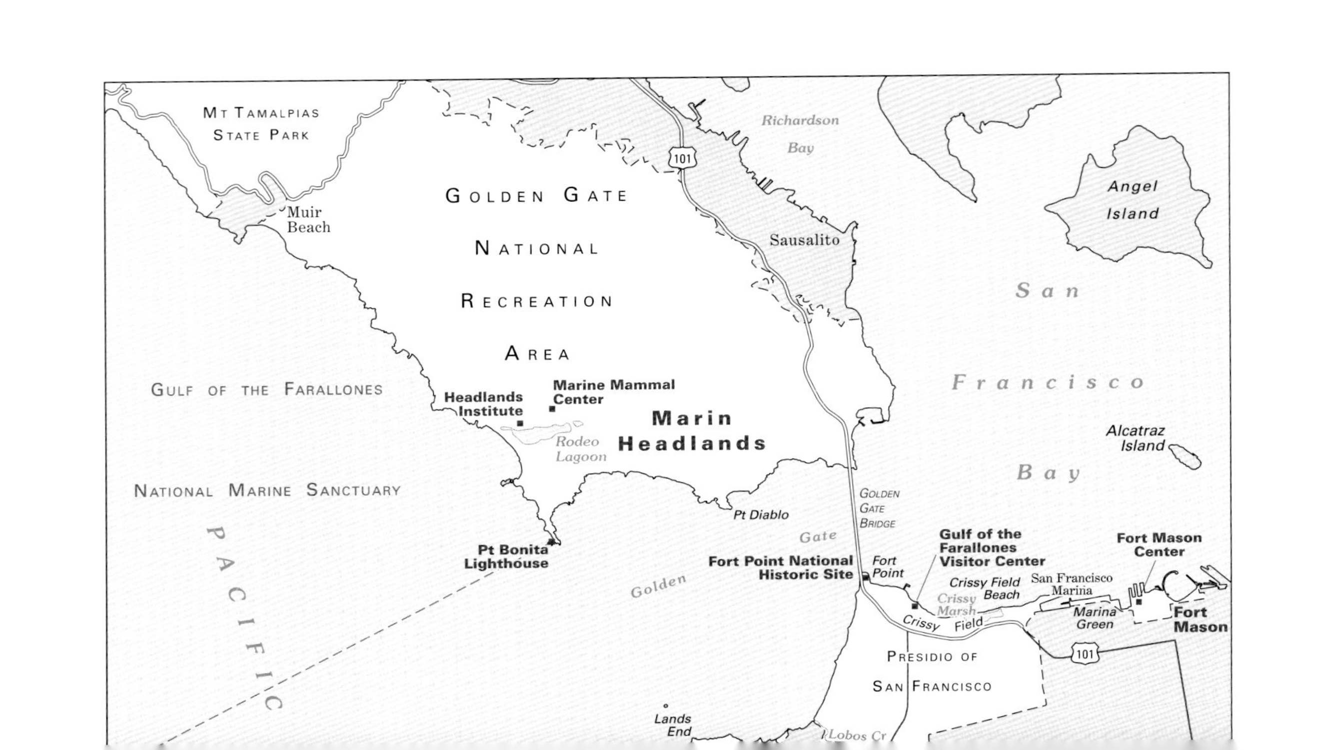

Mt Tamalpias
State Park
Muir
Beach
Golden Gate
National
Recreation
Area
101
Richardson
Bay
Sausalito
Angel
Island
San
Francisco
Bay
Alcatraz
Island
Gulf of the Farallones
National Marine Sanctuary
Headlands
Institute
Marine Mammal
Center
Marin
Headlands
Rodeo
Lagoon
Pt Bonita
Lighthouse
Pt Diablo
Golden
Gate
Golden
Gate
Bridge
Fort Point National
Historic Site
Fort
Point
Gulf of the
Farallones
Visitor Center
Crissy Field
Beach
Crissy
Marsh
Crissy
Field
San Francisco
Marina
Marina
Green
Fort Mason
Center
Fort
Mason
101
Presidio of
San Francisco
Pacific
Lands
End
Lobos Cr

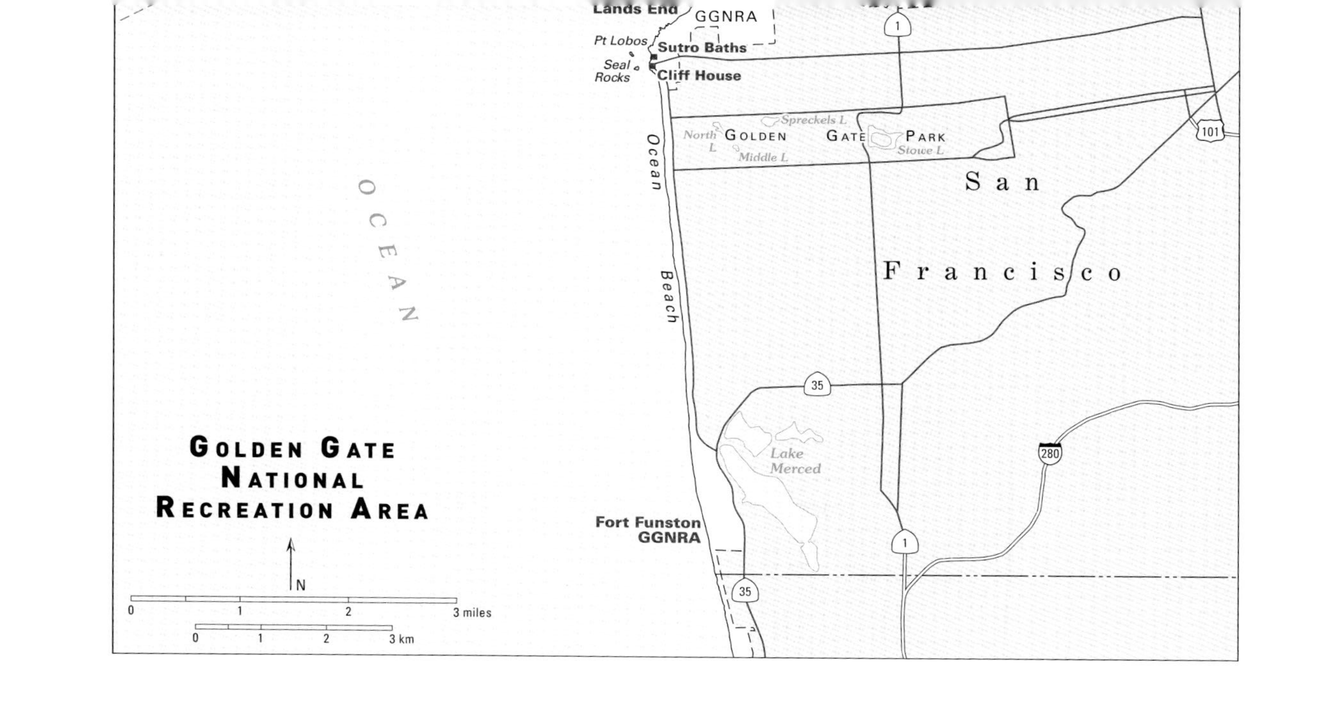
Lands End
GGNRA
Pt Lobos
Sutro Baths
Seal Rocks
Cliff House
1
North L
Golden Gate Park
Spreckels L
Middle L
Stowe L
Ocean Beach
San Francisco
101
OCEAN
35
Lake Merced
280
Golden Gate National Recreation Area
Fort Funston GGNRA
N
0
1
2
3 miles
3 km

North Lakes. Any of the forest corridors, flowering *Myoproum* bushes, or blackberry brambles offer opportunities for finding land birds. Other sites include Spreckles Lake and Lloyd Lake.

The California Academy of Sciences

55 MUSIC CONCOURSE DRIVE, GOLDEN GATE PARK, SAN FRANCISCO, CA 94118
(415) 379-8000
WWW.CALACADEMY.ORG
REGULARLY OPEN MONDAY–SATURDAY 9:30–5:00, SUNDAY 11–5

Cal Academy is a highly recommended stop for nature lovers. Like the Monterey Bay Aquarium, this is a world-renowned facility that requires at least a half day to explore fully. Exhibits include the Steinhart Aquarium, Morrison Planetarium, Rainforests of the World, and the Kimball Natural History Museum, which houses one of the most extensive collections of bird specimens (20 million plus) in the country. The blue whale skeleton suspended over *Tyrannosaurus rex* is in itself worthy of a visit. There is also an enthralling exhibit of live African Penguins.

Golden Gate National Recreation Area (GGNRA)

FORT MASON, BUILDING 201, SAN FRANCISCO, CA 94123-1307
(415) 561-4700
WWW.NPS.GOV/GOGA/INDEX.HTM

GGNRA is huge, encompassing a 60 mi swath of coastal lands from southern San Mateo County northward to Point Reyes. Within San Francisco, the bird-worthy spots are Fort Funston, the Presidio lands, and Crissy Field, including its shoreline and lagoon. Alcatraz Island is also a worthy stop during the seabird nesting season (March–June) but involves a full day and often crowds of tourists.

FORT FUNSTON This is a historic site on 230 acres of coastal bluffs at the end of Ocean Beach in southwesternmost San Francisco. The primary attraction for birders is the active Bank Swallow *(Riparia riparia)* colony that occupies the sandstone cliff face in spring and summer, one of only a few coastal California nesting sites. The Fort Funston bluffs provide these birds with two critical habitat components: friable soil, loose enough to burrow in but stable enough to last the nesting season; and

FORT FUNSTON'S BANK SWALLOW COLONY

To visit the Bank Swallow colony at Fort Funston, park in the southernmost parking area along the Great Hwy and walk south along the beach. Another option: Park in the lot near the intersection of Skyline and John Muir Drive. Follow the horse trail north along the ridge and down to the beach at the Great Hwy. From the main parking lot off Skyline Blvd, take the Sunset Trail or the horse trail north through the dunes past Battery Davis, then connect to the horse trail. Fort Funston is also accessible by Muni Bus 18.

foraging habitat over freshwater. Lake Merced, just to the east, provides the foraging habitat and is a good place to see the swallows in flight. Lake Merced is also one of the few coastal nesting sites for Clark's Grebe.

THE PRESIDIO LANDS These are extensive lands (about 800 acres) covering the northern edge of the city from Lobos Creek around Fort Point and the Golden Gate Bridge to Crissy Field. The 25 mi of hiking trails access a variety of habitats and wildlife viewing areas. The serpentine geology supports a unique coastal scrub community and several rare and endemic plants (e.g., Presidio clarkia, *Clarkia franciscana*). Other habitats include coastal bluff scrub, coastal dune scrub, salt-water marsh, freshwater marsh, arroyo willow, riparian forest, live oak forest, and extensive stands of Monterey cypress.

In the western side of the park, Lobos Creek Valley Trail (0.8 mi) is one excellent birding spot. The path and boardwalk wend through lupine shrub, grassland, and surrounding forest. This is one of the last areas where California Quail *(Callipepla californica)* have been sighted, now nearly extirpated from the city. However, some likely species include the Pygmy Nuthatch *(Sitta pygmaea)*, Allen's Hummingbird *(Selasphorus sasin)* from March to August, "Nuttall's" White-crowned Sparrow *(Zonotrichia leucophrys nuttalli)*, California *(Pipilo crissalis)* and Spotted *(P. maculates)* Towhees, and Bullock's Oriole *(Icterus bullockii)* in spring and summer, and this is one of the few places in the city where Western Bluebirds *(Sialia mexicana)* nest.

The Batteries to Bluff Trail (0.7 mi) provides a scenic path above the coastal bluffs with views of the "potato patch," a rough patch of ocean waters where currents collide and seabirds feed just outside the Golden Gate, a good place to scan the ocean for scoters, cormorants, Common Murres, and gulls.

CRISSY FIELD East of the Golden Gate, Crissy Field is set along the bay shore, with beach and dune habitats. The large lagoon attracts an ever-changing variety of waterbirds—loons and grebes, pelicans, cormorants, terns, diving and dabbling ducks—and, in late summer through fall, migrant shorebirds. Many rarities have been reported from the Crissy Field lagoon. The Gulf of the Farallones National Marine Sanctuary visitor's center is located within this portion of the park, and Western Snowy Plovers visit the beach a short walk to the north.

Marin County

Golden Gate National Recreation Area extends across the Golden Gate, into Marin County.

Marin Headlands

Perched above the north side of the Golden Gate is the premier hawk-watching site on the California coast. From August into October, raptors migrating southward along the Coast Ranges funnel toward the Golden Gate to make the shortest crossing over the waters of San Francisco Bay. The Golden Gate Raptor Observatory has been monitoring this epic passage for over 20 years. On a clear day with little wind, especially in late September, 500 or more raptors of 10 or more species can be seen passing over Hawk Hill at midday. Accipiters (Sharp-shinned and Cooper's Hawks) are the most common species by far, followed by Red-tailed Hawks and Turkey Vultures, but coastal rarities such as Broad-winged Hawks are surprisingly regular here, as are Peregrine Falcons and Merlins. Nineteen raptor species have been recorded here so far, so even Ferruginous Hawks, Swainson's Hawks, and an occasional eagle are possible.

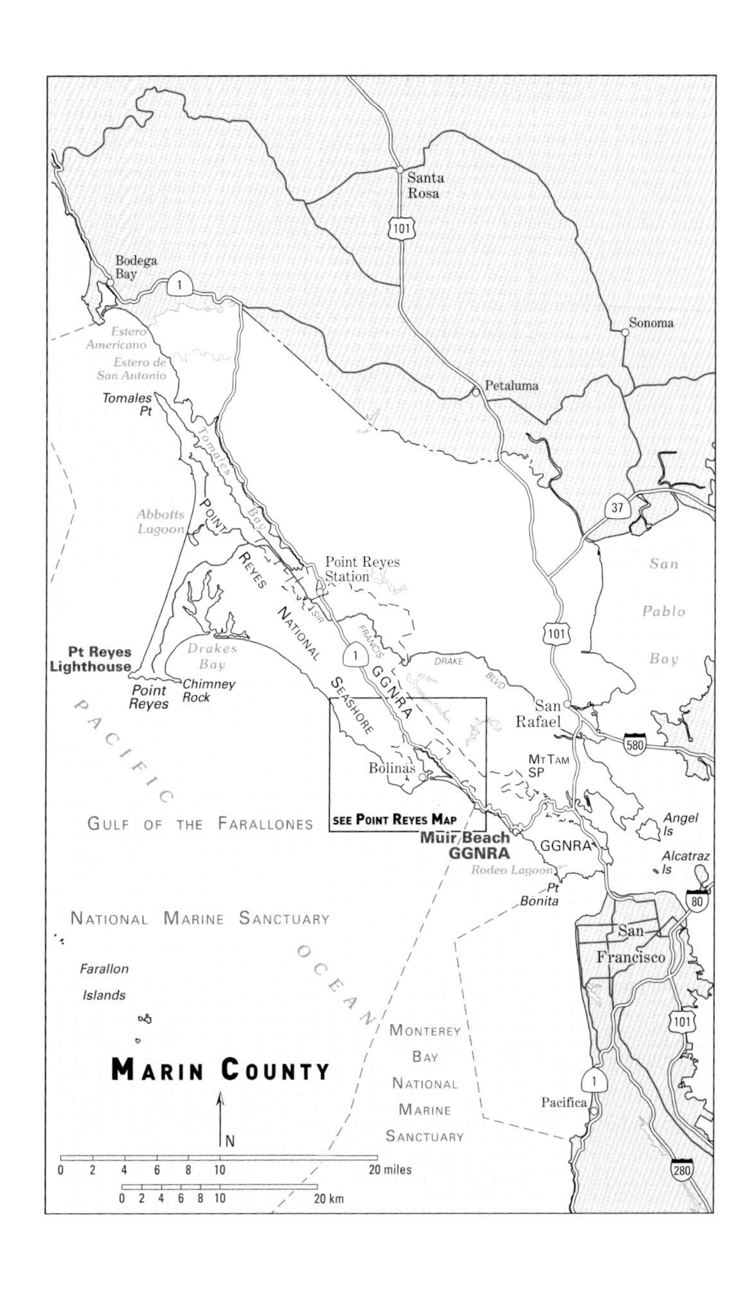

Santa Rosa
101
Bodega Bay
1
Sonoma
Estero Americano
Estero de San Antonio
Petaluma
Tomales Pt
Tomales Bay
37
Abbotts Lagoon
POINT REYES NATIONAL SEASHORE
San Pablo Bay
Point Reyes Station
JSIR
FRANCIS
101
Pt Reyes Lighthouse
Drakes Bay
1
GGNRA
DRAKE BLVD
Point Reyes
Chimney Rock
San Rafael
PACIFIC
580
Mt Tam SP
Bolinas
GULF OF THE FARALLONES
SEE POINT REYES MAP
Angel Is
Muir Beach GGNRA
GGNRA
Alcatraz Is
Rodeo Lagoon
Pt Bonita
80
NATIONAL MARINE SANCTUARY
San Francisco
OCEAN
Farallon Islands
101
MONTEREY BAY NATIONAL MARINE SANCTUARY
MARIN COUNTY
1
Pacifica
N
0 2 4 6 8 10 20 miles
0 2 4 6 8 10 20 km
280

VISITING HAWK HILL AT MARIN HEADLANDS

From the south Cross the Golden Gate Bridge and take the Alexander Avenue (second) exit. Get into the left lane and follow signs to "101 San Francisco." At the stop sign, turn left and go under the freeway. Then turn right on Conzelman Road going toward Hawk Hill, drive 1.8 mi, and park at the top of the hill off the road before it becomes one-way. Walk up the west side of Hawk Hill, pass the locked gate, and walk just a few hundred more feet to reach the top of Hawk Hill.

From the north After exiting the Waldo Tunnel, take the last Sausalito exit to the Golden Gate National Recreation Area (follow signs to "Golden Gate National Recreation Area / Sausalito"). This exit ramp is very short and ends in a stop sign. You need to be on the left. Turn left at the stop sign, and then turn right immediately onto Conzelman Road and go up the hill toward Hawk Hill. Drive 1.8 mi, and park at the top of the hill off the road before it becomes one way. Walk up the west side of Hawk Hill, pass the locked gate, and walk just a few hundred more feet to reach the top of Hawk Hill.

Tiburon Audubon Center of the National Audubon Society

376 GREENWOOD BEACH ROAD, TIBURON, CA 94920
(415) 388-2524; RICHARDSONBAYCENTER@AUDUBON.ORG
WWW.TIBURONAUDUBON.ORG
OPEN MONDAY–FRIDAY 9–5 ALL YEAR.
AMENITIES INCLUDE A VISITOR'S CENTER, SMALL BUT WELL-STOCKED NATURE BOOKSTORE, BATHROOMS, WATER, PICNIC TABLES, DISPLAYS, NATURE WALKS, AND PERIODIC TOURS OF THE HISTORIC LYFORD HOUSE.

Tiburon is located 1 mi east of Hwy 101 along Tiburon Blvd, north of Marin City and south of Corte Madera. From either direction on Hwy 101, take the Tiburon Blvd exit 1 mi east to the traffic signal at Greenwood Cove Drive. Turn right, follow the left swerve, and go up the little hill to the sanctuary entrance. (This is where Greenwood Cove Drive becomes Greenwood Beach Road.)

The Tiburon center is a 900-acre waterbird sanctuary on San Francisco Bay with easy trails to bird-viewing sites and the historically significant Victorian Lyford House. The friendly and knowledgeable staff will provide answers to your questions. Blackie's Pasture is just outside the north boundary of the preserve.

There is a full calendar of events (two to six per month), including slideshow lectures and field trips listed in the Audubon Center quarterly newsletter. Specifically for young people, the Summer Audubon Adventure weekly camps run from June into August, Monday through Friday, and there are docent programs from Junior Nature Guides. Call the center for a schedule of naturalist-led walks at the sanctuary.

During winter months, Richardson Bay, which is nested within central San Francisco Bay, supports an abundance of waterbirds that are easily observed through binoculars or telescopes from overlooks along the sanctuary's bluff or from shore. Grebes, scaup, Buffleheads, goldeneyes, Surf Scoters, various gulls, and Double-crested Cormorants are especially numerous (in the thousands) when a run of Pacific Herring *(Culpea pallasii)* enters the bay to spawn, usually in November or December. Richardson Bay is one of the principle spawning sites for Pacific Herring on the Pacific coast and has been designated an Important Bird Area by the National Audubon Society for the abundance of birds it attracts annually. The birds forage on adult fish, but more so on the herring eggs that the fish deposit on submerged aquatic vegetation (algae and eelgrass). The eggs are laid in the millions, but the huge rafts of birds consuming the bounty make it a wonder that any ever hatch to continue the herring population. In March and April it is possible to see the amazing courtship dances of Clark's and Western Grebes.

When waterbirds are scarce, late April through September, many land bird species may be closely observed in the riparian habitat at the pond or frequenting the feeders at the nearby visitor's center.

Audubon Canyon Ranch (ACR)/ Bolinas Lagoon Preserve

4900 HWY 1, STINSON BEACH, CA 94970

(415) 868-9244; ACR@EGRET.ORG

WWW.EGRET.ORG

AMENITIES INCLUDE VISITOR'S CENTER DISPLAYS AND NATURE BOOKSTORE, WHEELCHAIR-ACCESSIBLE BIRD BLIND AND PICNICKING, WATER, BATHROOMS, AND SHORT TRAILS, INCLUDING ONE TO THE ROOKERY OVERLOOK. DOCENTS (RANCH GUIDES) WILL TELL YOU FACTS OF NATURE AND ANSWER QUESTIONS.

OPEN ONLY FROM THE SECOND WEEKEND IN MARCH THROUGH THE SECOND WEEKEND IN JULY.

One way to get to ACR is to take Hwy 101 north from the Golden Gate or south from San Rafael to the exit for Sir Francis Drake Blvd. Follow that, through several connected small towns, then parkland, for 21 mi to Olema. Turn south (left) on Hwy 1, and travel through the beautiful Olema Valley south about 10 mi to Bolinas Lagoon. Stay on Hwy 1, and ACR is on the left in about 1 mi.

Another way to get there is to take Hwy 101 north from the Golden Gate or south from San Rafael to the Hwy 1 / Stinson Beach exit. Continue about 12 (curvy) mi to the town of Stinson Beach (this route goes very near to the entrance of Muir Woods National Monument), then 3.5 mi north to the ACR preserve, on your right.

During the breeding season (mid-March to mid-July) herons and egrets nest high in a grove of Coast Redwoods. The Rawling's Trail leads up the canyon to an overlook that affords excellent views of the nesting colony from above.

Each year over 100 ACR docents offer free programs to thousands of fourth- and fifth-grade school children. There are free workshops for selected teachers, and packets of teaching materials. Docents (two) will visit classrooms and present a slide show and hands-on activities and materials. At the preserve, one docent will lead five to seven students on a three- to four-hour nature walk through the display halls and outside, through several different wildlife habitats.

NEARBY BIRDING ACR is adjacent to Bolinas Lagoon, which provides excellent birding throughout the year but especially from September through April. Many turnouts on the lagoon

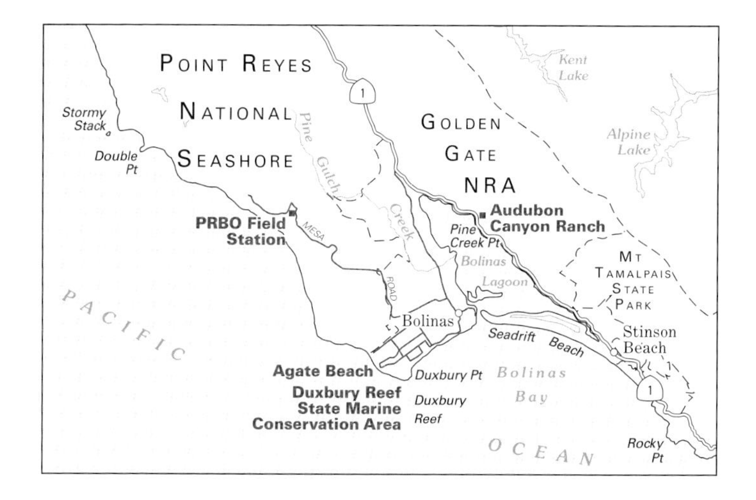

side of Hwy 1 (most safely accessed going north to south) bring you very close to the water and (at lower tides) mudflats.

The lagoon's habitats are crowded with grebes and ducks or plovers and sandpipers of many species September through April. During summer and early fall, huge numbers of Brown Pelicans, Heermann's Gulls, and Elegant Terns ("the three amigos") use mud bars for preening and resting. Ospreys are almost always within view, and the chick-fledging activities of the herons and egrets are entertaining and sometimes comical. The songs of many sorts of land birds ring loudly from the oak-fir forests and riparian cover along streams on canyon floors.

Palomarin Field Station of Point Reyes Bird Observatory (PRBO) Conservation Science

MESA ROAD, BOLINAS, CA 94924

(415) 868-0655

WWW.PRBO.ORG/CMS/27

OPEN MAY–THANKSGIVING, SUNRISE TO NOON, EVERY DAY BUT MONDAY; THANKSGIVING–MAY, SUNRISE TO NOON ON WEDNESDAY, SATURDAY, AND SUNDAY. THE UNSTAFFED VISITOR CENTER IS OPEN YEAR-ROUND SUNRISE TO SUNSET.

At the north end of Bolinas Lagoon, there is an unmarked turn-off to the left. Turn left here, go one block and turn left again.

Follow this road around the other side of the lagoon toward Bolinas. Turn left at the first stop sign. The second stop sign will be Mesa Road. Take a right onto Mesa Road (see green road sign "Point Reyes Bird Obs. 4 Mi"). Point Reyes Bird Observatory is 4 mi down Mesa Road. The road will become a dirt road at about 3.25 mi. Follow the dirt road until you see the sign for PRBO on your left. Turn in at the dirt driveway on your left.

The field station is very near Audubon Canyon Ranch, and both can be visited on the same day, but it's best to visit Palomarin in the early morning when land birds are most active.

PRBO Conservation Science/Palomarin Field Station

MAILING ADDRESS: PRBO, 4990 SHORELINE HWY, STINSON BEACH, CA 94970

(415) 868-0655 PALOMARIN HOTLINE EXTENSION 395, (415) 868-0655 GROUP SCHEDULING EXTENSION 315

WWW.PRBO.ORG; PALOMARIN LINK: WWW.PRBO.ORG/PALO

This is an avian research facility and bird-banding laboratory of PRBO Conservation Science, also known as Point Blue Conservation Science. There is a visitor's center and small museum that is wheelchair accessible and always open. A nature trail with interpretative signs and roads or trails wends its way through coastal scrub and forest.

LEARNING Bird banding takes place (dependent on weather) from May 1 to Thanksgiving every day except Monday, dawn until noon. From Thanksgiving until the end of April, banding happens weekends and Wednesdays, dawn until noon. Staff and/or interns will welcome you and explain what is going on.

Groups or classes may schedule appointments at (415) 868-0655 extension 315. The Palomarin Hotline, (415) 868-0655 extension 395, will give you recent updates.

NEARBY BIRDING The Arroyo Hondo Trail (1.5 mi round trip) is near Palomarin and is especially rewarding April through September. It borders a stream and ascends through several habitats into old-growth Douglas fir. Back at Hwy 1, Bolinas Lagoon provides excellent waterbird birding throughout the year—especially September through April (see entry for Audubon Canyon Ranch, above), and the several canyons accessible from Hwy 1 on the east shore are great for land birds, especially mid-March through October.

Point Reyes National Seashore

BEAR VALLEY VISITOR'S CENTER, 1 BEAR VALLEY ROAD, POINT REYES STATION, CA 94956

(415) 464-5100 PARK INFORMATION, (415) 669-1534 POINT REYES LIGHTHOUSE

OPEN THURSDAY–MONDAY 10–4:30

WWW.NPS.GOV/PORE/INDEX.HTM

The Point Reyes Peninsula covers 100 sq mi of wildlands, including over 50 mi of dramatic rocky shoreline and headlands, sweeping beaches and dunes, wildlife-wealthy lagoons and estuaries, tidal marshes, extensive conifer forests, northern coastal scrub, coastal prairie, grasslands, and over 150 mi of hiking trails. All this habitat diversity, coupled with an equable climate and serendipitous geographical setting, make Point Reyes a birding hot spot.

The peninsula's wildlife checklists include over 480 species of birds, 90 mammals, 24 amphibians and reptiles, and nearly 1,000 plant species, not to mention innumerable invertebrates. The unique geology of the peninsula, wholly distinct from the adjacent mainland and defined by the San Andreas rift zone, adds to the peninsula's distinctive character.

Birding the seashore and vicinity can take a day, a week, or a lifetime. As with all other sites mentioned in this book, season is an important consideration. Following are some of the primary sites to visit, followed by seasonal suggestions.

OUTER POINT The Outer Point includes the Point Reyes Lighthouse, Chimney Rock, the Fish Docks, and Drakes Beach, as well as islands of vegetation surrounding the ranches scattered along Sir Francis Drake Blvd to the lighthouse. During spring and summer, the nesting colonies of seabirds are best viewed from overlooks at the outermost headlands, the lighthouse, Chimney Rock, and the sea lion overlook, if the fog is not too thick! Common nesting species include Common Murre, three species of cormorants, and Western Gull. Less common are Pigeon Guillemot and Black Oystercatcher. Rarest, but present, are Tufted Puffin and Ashy Storm-Petrel. (The Common Murre colony at Point Reyes, recovering steadily from declines related to oil spills and other disturbances in the 1980s, hosts about 40,000 individuals, best viewed from the lighthouse.) During fall migration, late August into early October, the Outer Point is visited by waves of migrant land birds, including unusual numbers of rare "vagrants." Migrant flocks are found in the islands of cypress trees (planted as windbreaks) at the various ranches, on the path to the lighthouse, or near the ranger station near the fish docks. On weekends, when the weather is favorable, migrant flocks are often pursued by throngs of eager birders. During the winter months, Gray Whale watching from the lighthouse is a popular pastime. The park offers shuttle buses from headquarters and from South Beach on weekends. Other colonies of marine mammals—Northern Elephant Seals *(Mirounga angustirostris)*, California Sea Lions *(Zalophus californianus)*, and Harbor Seals *(Phoca vitulina)*—are viewable from designated overlooks within the park. Peregrines nest in the headland cliffs and often power fly past, at eye level.

From fall through spring, Point Reyes's tidal wetlands—Bolinas Lagoon, Limantour and Drake's Esteros, Abbott's Lagoon, and Tomales Bay—host vast numbers of waterbirds. Diving and dabbling ducks, loons and grebes, pelicans and cormorants, egrets and herons, and 20 or so species of shorebirds occur regularly. Numbers peak from November through February, but interesting birds can be seen any time of year. During spring and fall migration, rarities may sometimes be found.

Freshwater habitats are much less common on the peninsula, but some pond-loving ducks, such as Wood Ducks and Gadwall, can usually be found at the pond near the Five Brooks trailhead or some of the artificial stock ponds scattered around the peninsula.

TOMALES BAY North of Point Reyes on Hwy 1 toward Sonoma County, the road parallels the east shore of Tomales Bay, designated as "a wetland of international importance" (RAMSAR site) because of its value as waterbird habitat. Diving ducks—especially Bufflehead, Greater Scaup, and Surf Scoter—are abundant in the bay in winter, and during herring runs they are joined by impressive numbers of Brandt's Cormorants, as well as loons and grebes. There are many pullouts on the shoulder of the highway that afford safe stops for scanning the bay waters and looking for unusual species amid the large rafts of more common waterbirds. The small town of Marshall is 10 mi north of Point Reyes Station, and another 3.4 mi north of Marshall is Miller County Park, which has a public parking area worth a brief stop. From here, Hog Island is visible in the middle of the bay. Off the north end of the island, submerged eelgrass beds provide spawning habitat for Pacific Herring and forage for Black Brant, which are present most months but peak during spring migration (April). A few miles to the north is a worthy overview above the mouth of Walker Creek. Here scan the bay shore for American White Pelicans, large shorebirds, and raptors.

Sonoma County

Bodega Bay

Because there are no bird-oriented environmental centers along the Sonoma coast, here are directions to the best birding localities.

BODEGA HARBOR Located just west of Hwy 1 along south coastal Sonoma County, Bodega Harbor provides excellent year-round birding, great scenery, and fresh seafood cuisine.

DORAN BEACH This is the giant sandbar that encloses the harbor's south side. There are signs to Doran Beach along Hwy 1, south of town before it turns inland. There is a $7 per vehicle fee to enter Doran Regional Park—a small price for what you might see.

For best birding results you will want to consult a daily tide log. Try to plan to be at Doran on a rising tide. Drive through the pay booth, go a few hundred yards and park in the paved lot enclosed by young Monterey cypress trees. There is a bathroom with running water here. Cross the road and walk northward to the Rich Stallcup Memorial Observation Deck for an overview of the harbor. As the tide moves in (this is a *shallow* estuary, so the water level rises quickly), it pushes shorebirds toward you from recently inundated foraging habitats. During fall migration, 15 species of shorebirds (fewer at other seasons) will cluster close to the shoreline, usually moving to the harbor's northeast corner at high tide. Marbled Godwits will be the most obvious birds, sometimes (except for summer) numbering in the low thousands. In addition to waders, gulls and terns gather on the tidal islands. Be sure to scan posts and "the box" in the salt marsh to the northeast for a perched harrier, Osprey, or Peregrine.

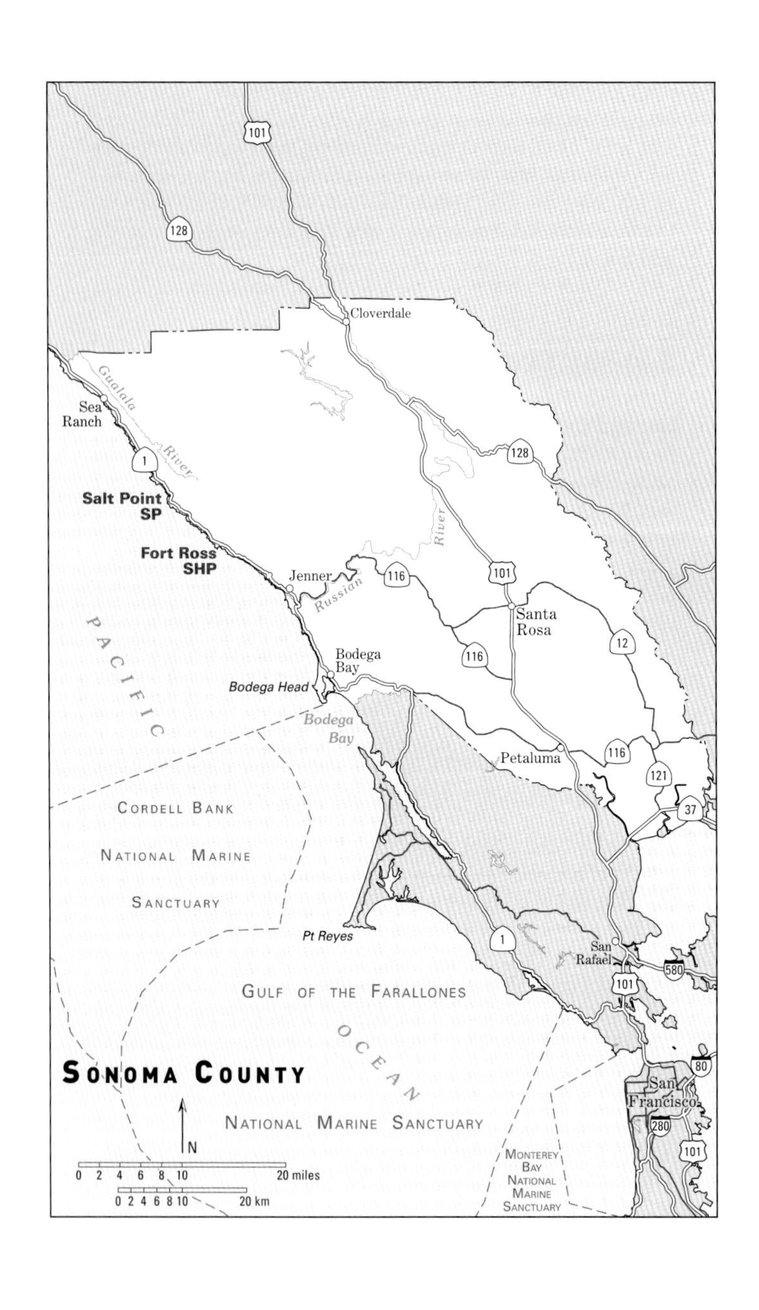
101
128
Cloverdale
Gualala
River
Sea
Ranch
1
128
Salt Point
SP
River
Fort Ross
SHP
Jenner
116
101
Russian
Santa
Rosa
PACIFIC
12
Bodega
Bay
116
Bodega Head
Bodega
Bay
Petaluma
116
121
37
CORDELL BANK
NATIONAL MARINE
SANCTUARY
Pt Reyes
1
San
Rafael
580
101
GULF OF THE FARALLONES
OCEAN
80
SONOMA COUNTY
San
Francisco
NATIONAL MARINE SANCTUARY
280
N
MONTEREY
BAY
NATIONAL
MARINE
SANCTUARY
101
0 2 4 6 8 10 20 miles
0 2 4 6 8 10 20 km

Across the road to the south is a stretch of protected sandy beach, a good area to look for Snowy Plovers and Sanderlings, and up to six species of gulls. Off the beach, out on the bay are various surface divers—loons, grebes, scoters, murres, and guillemots—mostly in winter. This is also a fairly good place to find Marbled Murrelets, often rather close to shore. Doran Beach Road ends at the mouth of the harbor. Habitats here are sandy beach, mudflat, rocky shore (the breakwaters), ocean, and bay; waterbirds are practically everywhere.

BODEGA HEAD Go back to Hwy 1 and head north, passing two worthy seafood eateries—Lucas Wharf and The Tides Restaurant (a movie set in Hitchcock's *The Birds*). After about 1 mi, turn left at the sign for Westside Park, Marina, and Bodega Head. The road is well signed and takes you downhill to a stop sign. Turn right here, on Bayflat Road, and drive around the west side of the harbor, all the way to the end at the open-ocean viewpoint on Bodega Head.

Exercise *extreme* caution here, as several people, enamored with the view, have slipped and fallen to death on the rocks below. The Head is excellent for viewing nesting species like Western Gull, Black Oystercatcher, Brandt's and Pelagic Cormorants, and Pigeon Guillemot in the spring and for observing rocky shore birds (turnstones, tattlers, Surfbirds, and oystercatchers) and Gray Whales in winter. Many kinds of true seabirds (shearwaters, fulmars, kittiwakes, auklets, and even albatross) have been identified from this point by patient observers with high-powered telescopes.

CAMPBELL COVE Go back down the hill toward the harbor entrance, which is easy to see. At the bottom, turn right into the parking lot at Campbell Cove. This spot, called "hole-in-the-head" by local birders, was once the planned location for a nuclear power plant—until someone pointed out that the San Andreas Fault runs directly below!

Again, from September through March there should be no difficulty locating many sorts of large, close waterbirds (and raptors) to study. A short boardwalk from the Campbell Cove parking lot ends at a small pool with sedge marsh and wax myrtle edges. There are often several Black-crowned Night-Herons partly concealed in the myrtles across the pond, a Pied-billed Grebe or two *on* the pond, and, in winter, Yellow-rumped Warblers (of the Myrtle subspecies) everywhere. Migrant land birds seem to be attracted to the vegetation surrounding the pond, especially in autumn.

BODEGA HARBOR There are many safe dirt turnouts on your way back to Hwy 1 where you will see loons, grebes, ducks, and Brant at high tide and myriad shorebirds and gulls at low tide. Watch for Ospreys, White and Brown Pelicans, and terns among the more usual suspects.

It is wise to spend a whole day or more in this small area called Bodega Harbor. Birds are abundant and close, and the species mix at any one spot is constantly changing, as are the elevations of the tide that dictates the behavior of life and mud.

JENNER Eleven miles north of Bodega Bay on Hwy 1 is the small coastal village of Jenner, located at the mouth of the Russian River. Just south of town, before crossing the river, Goat Rock Road leads to Goat Rock Beach State Park, worth checking for rocky shore birds and diving birds nearshore. A mile beyond town, on the west side, is a large pullout that affords a good overview of the river mouth and the sandbar, where large numbers of Harbor Seals haul out, usually accompanied by flocks of gulls. Scan the surf line for scoters and loons.

FORT ROSS STATE HISTORIC PARK Heading north from Jenner, it's another 47 mi of windy coastal highway to Point Arena in northernmost Sonoma County. En route you'll pass Fort Ross State Historic Park, the southernmost Russian settlement on the West Coast, established in the early 1800s on a Pomo village known as *Metini*. The Russian outpost was established as a base for hunting marine mammals, particularly Sea Otters, their luxuriant pelts being highly prized by the Chinese Mandarins for their warmth and beauty. Sadly, Sea Otters are no longer resident along the Sonoma coast, having been hunted to oblivion. The park encompasses over 3,000 acres of coastal habitats, including redwood forest, coastal grasslands, sandy beach, and rocky shore.

SALT POINT STATE PARK This park 6.5 mi north of Fort Point has camping facilities and includes an even more diverse mix of coastal habitats, including a pygmy forest of cypress, pine, and redwood. Gerstle Cove Marine Reserve at Salt Point is one of the first underwater parks in California, where marine life is completely protected. Tide-pooling here is excellent.

GUALALA POINT REGIONAL PARK In northernmost Sonoma County, 36 mi north of Jenner and 1 mi south of the town of

Gualala (at mile marker 58.50), Gualala Point Regional Park offers public access to the mouth of the Gualala River. This is one of the county's premier birding spots. Common Mergansers are often in the river, and just offshore you can expect to find loons, grebes, cormorants, scoters, and pelicans most months. During the nesting season, a nice variety of alcids may be found here as well—Common Murre, Pigeon Guillemot, Marbled Murrelet, and Rhinoceros Auklet. About 4 mi to the north, offshore from Anchor Bay (Mendocino Co.) are Fish Rocks, a seabird nesting site quite close to shore, which supports a few pairs of Tufted Puffins.

Mendocino County

A few bird-oriented centers are listed below, but first here are directions to some birding localities along the coast not connected to centers.

ARENA COVE On the west side of the town of Point Arena is Arena Cove, a good place to check for marine birds. From the pier at the end of Port Road you may see "Al," the Laysan Albatross that spent at least 17 consecutive winters in this cove. Interestingly, this is the closest spot on the North American continent to the Hawaiian Islands, the main breeding area for this species of albatross.

POINT ARENA LIGHTHOUSE Approximately 5 mi north of Point Arena, the Garcia River meets the Pacific. The bluffs, marshes, and pastureland in the immediate vicinity of the river mouth are worth checking; the buffs along the road to the Point Arena Lighthouse afford an especially good view of the river's mouth. The property outside the gates of the lighthouse was previously privately owned but is now public, managed by the Bureau of Land Management.

About 0.5 mi north of Lighthouse Road on Hwy 1 is Miner Hole Road, a dirt road that is very overgrown. It is riddled with potholes and ruts, and you will need a vehicle with some clearance to drive down it. From here, the Garcia River estuary is visible to the south. At the end of the road you can park and make your way down to the river.

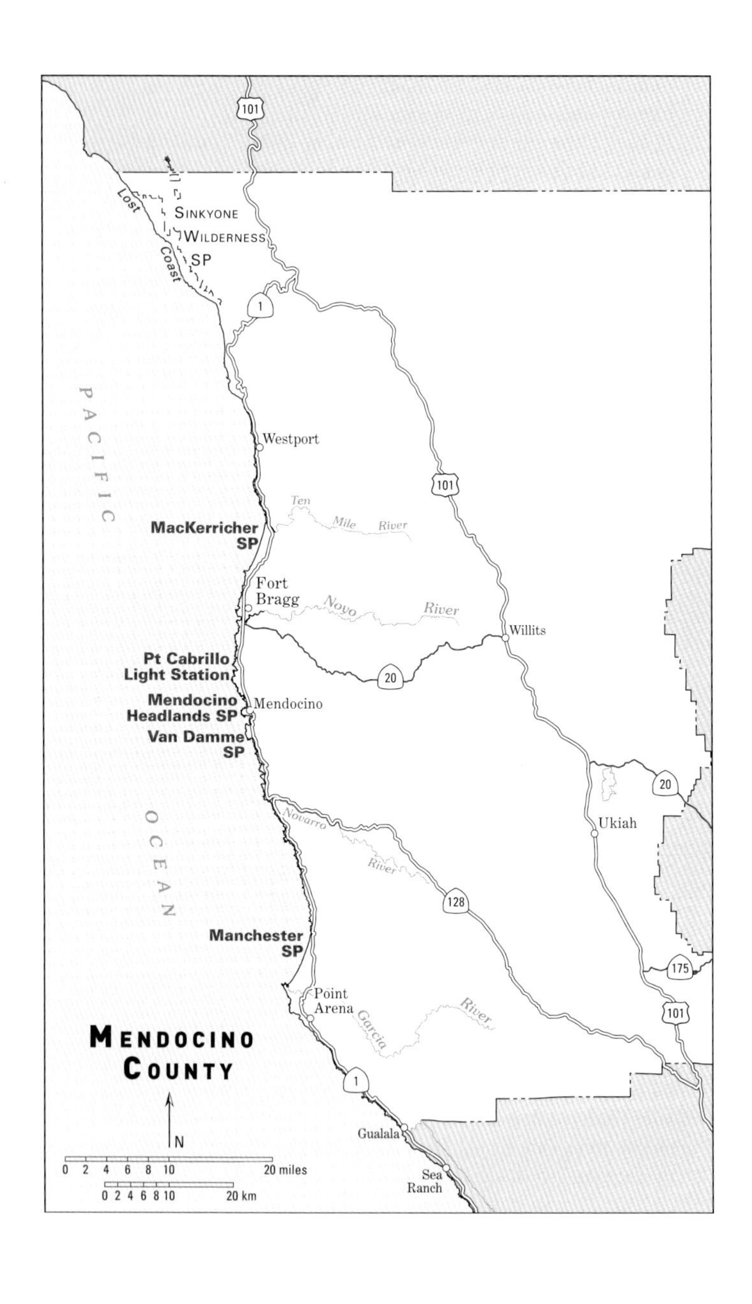

101
Lost
Coast
SINKYONE
WILDERNESS
SP
1
PACIFIC
OCEAN
Westport
101
Ten
Mile
River
MacKerricher
SP
Fort
Bragg
Novo
River
Willits
Pt Cabrillo
Light Station
20
Mendocino
Headlands SP
Mendocino
Van Damme
SP
20
Novarro
River
Ukiah
128
Manchester
SP
175
Point
Arena
Garcia
River
101
MENDOCINO
COUNTY
1
N
Gualala
0 2 4 6 8 10 20 miles
0 2 4 6 8 10 20 km
Sea
Ranch

GARCIA RIVER FLATS Hwy 1 crosses the bottomlands, just to the east of the mouth, and allow views of the Garcia River Flats. Sizable flocks of Tundra Swans graze and roost in these low pasturelands from November through April. Most swans winter in the Central Valley; the Garcia River flats is one of the few, if not the only, California coastal sites where substantial numbers are reliable in winter. In addition to the swans, other birds such as teal (three species), Mallards, Marbled Godwit, Canada Goose, Brant, and various species of gulls also utilize this pastureland. Another important area near the river for waterbirds is a small freshwater marsh east of Hwy 1 and north of the river. Dabbling ducks such as Pintail, Gadwall, Mallard, and Cinnamon Teal can be found in this wetland.

MENDOCINO COAST IMPORTANT BIRD AREA The Mendocino Coast Important Bird Area (IBA) is identified by the National Audubon Society as a 20 mi stretch of coast from Mendocino Headlands north to the mouth of the Ten Mile River. The area encompasses a diverse habitat mix—northern coastal prairie and scrub, grassland, coniferous forest, riparian corridors, freshwater wetlands, and brackish lagoons. Much of the area is protected and open to the public, from south to north: Van Damme State Park, Mendocino Headlands State Park, Russian Gulch State Park, and MacKerricher State Park. Park information is available from (800) 777-0369 or www.parks.ca.gov.

Van Damme State Park (1,831 acres)

(707) 937-5804

The park is located 3 mi south of the town of Mendocino on Hwy 1. The highway runs through the park, separating the campground and the Fern Canyon trailhead to the east and the beach and parking lot to the west. Little River runs through Fern Canyon and offers some luxuriant riparian habitat for land-birding. The offshore rocks host significant seabird nesting colonies, visible from shore. The islets in Van Damme Cove may be the farthest-north known nesting site of the highly pelagic Ashy Storm-Petrels, but they visit their burrows only under the cloak of darkness and the chances of viewing them are negligible. Nevertheless, it's comforting to know they are in the neighborhood. Camping is available at Van Damme, and a specialty around the campgrounds is the Gray Jay *(Perisoreus canadensis)*, relatively rare on the California coast.

Mendocino Headlands State Park

From the town of Mendocino, take Main Street north past the Mendocino Hotel to Heeser Drive and park. A trail heads southwest, through a stand of bishop's pine and then descends to the mouth of the Big River. The Big River estuary, upstream from the mouth, attracts waterbirds, especially in winter.

The islets offshore are very close, providing for excellent observation of roosting and nesting seabirds. Goat Island is just offshore, and between mid-April and late August, nesting Brandt's and Pelagic Cormorants can easily be viewed, while equivalent numbers of Common Murres nest on the outer ends of the northwest islets. Black Oystercatchers also nest in the area. Small numbers of Rhinoceros Auklets and Tufted Puffins nest here as well, though they present more of a viewing challenge.

Russian Gulch State Park

(707) 937-5397

CAMPING IS AVAILABLE.

Just north of the town of Mendocino, Russian Gulch Creek Canyon meets the roily Pacific, presenting a dramatic coastal landscape. The dense forests here reach down to the craggy coastal bluffs. A deep cave at tidal level known as the Devil's Punchbowl is a blowhole accessible from the Headlands Trail and is worth seeing—and hearing—when the waves roll in at high tide, echoing through the cavernous geology. Spanning more than a mile of coastal headlands, the park supports extensive habitat for rocky shore birds—turnstones, tattlers, and oystercatchers—and nearshore marine species.

Point Cabrillo Light Station and Preserve

45300 LIGHTHOUSE ROAD, MENDOCINO, CA 95437

(800) 262-7801

The Light is a California Historic Park located just 1.5 mi south of Fort Bragg and an ideal ocean overlook for marine birds and whales. Pelagic Cormorants, nesting in the cliff faces, are easily viewed here in spring and summer.

Fort Bragg

Fort Bragg is the largest town on the Mendocino coast, a good place to refuel, find lunch, or visit some art studios. (Mendocino

Coast Photographer Guild and Gallery, 357 North Franklin Street, specializes in photography of wildlife, especially birds, and natural landscapes.)

Glass Beach, right in the town of Fort Bragg, can be a "quick stop." This beach is a former dump and gets its name from tide-tumbled pieces of colored glass. During the winter, it is also the best place in California to view the resplendent Harlequin Duck. From Hwy 1 in north Fort Bragg, turn west on Elm Street and drive 0.2 mi to the end and park. Walk on along the dirt path toward the ocean. Along the way you might see several species of sparrows, a Black Phoebe, or a White-tailed Kite perched or hovering off to the right. Beyond the gate, about 0.2 mi, search the roiling water and dry rocks to the left for Harlequin Ducks in winter. More usual cormorants, scoters, and grebes may be around at high tide. From November through March look for roosting rocky shore birds on dry islets off to the right.

MacKerricher State Park

24100 MACKERRICHER ROAD, FORT BRAGG, CA 95437

(707) 937-5804

The park extends approximately nine miles along the coast. The shoreline of its south half consists of rocky head-lands separated by sandy beaches and coves. Miles of gently sloping beach lie in the northern half.

VIRGINIA CREEK BEACH This beach is about 1.5 mi from Fort Bragg, a popular spot for beachgoers but also nesting grounds for the rare Western Snowy Plover.

LAKE CLEONE AND THE LAGUNA POINT BOARDWALK Three mi north of Fort Bragg, the Mill Creek Drive exit from Hwy 1 takes you to the entrance road. This road will take you past three campgrounds, Lake Cleone, and the Laguna Point boardwalk. These are beautiful places and if the weather is good, there are excitingly productive hikes and pleasant picnicking opportunities. Lake Cleone—a former tidal lagoon, now a freshwater lake—starts at the east end of the parking lot, the starting point for a nice, woodsy walk around the 30-acre lake. The boardwalk onto the headlands at Laguna Point may be excellent for birding (mostly October through April) and seal and whale watching. From different spots (there are viewing platforms) along the Laguna Point trail, many species of coastal birds (loons, grebes, cormorants, scoters, and alcids) may be seen well through binoculars and telescopes, and focusing a bit farther, toward the horizon, may reveal true seabirds such as Northern Fulmars or Black-legged Kittiwakes.

The sand dunes at the north end of MacKerricher extend for nearly 5 mi, paralleling the shoreline, perhaps the most pristine coastal dune system in California, and certainly the most protected. The dunes end at the mouth of the Ten Mile River, the northern reach of the Mendocino IBA. The dune area is one of the most reliable places on the coast to find Common Nighthawk *(Chordeiles minor)* hawking insects on the wing on a summer evening. (Nighthawks are very rare on the coast south of Mendocino Co., another mysterious aspect of this enigmatic bird.) The dunes system also includes a rare feature known as a "fen-carr" ecosystem, a waterlogged bog supporting riparian vegetation, apparently the only example in California (Warner and Hendrix 1984). Tidal marshlands at the mouth of the Ten Mile River and the associated brackish lagoon are bird-friendly spots, and the river supports a remnant run of threatened Coho salmon and the endangered Tidewater Goby *(Eucyclogobius newberryi).* As with the storm-petrels at Van Damme State Park, you are not likely to see the gobies, but it is reassuring to know they are there.

Humboldt County

Southern Humboldt County

Humboldt County is big, and the southern quarter—40 mi or so of coastline, from Point Delgado to Cape Mendocino—is remote and largely inaccessible except to the most intrepid traveler, hence the region (including northern Mendocino Co.) is aptly named the Lost Coast.

The approximate center of Humboldt's coastline is between the mouths of the Eel River and the Mad River, which bracket Humboldt Bay, one of the West Coast's premier estuaries. The Humboldt Bay estuary includes the deltaic wetlands of three major regional watercourses—the Eel, Elk, and Mad Rivers. The bottomlands of the Eel River, largely west of Hwy 101, encompass a maze of sloughs and islands, largely inaccessible from land.

LOLETA BOTTOMS In the south bay area, excellent birding opportunities can be found in the Loleta Bottoms. The town of Loleta is 13 mi south of Eureka. Wend through the town and find Cannibal Island Road; head west through the bottoms toward Crab Park, scanning the pastures for raptors, plovers and other shorebirds, and in fall and winter, longspurs.

HUMBOLDT BAY The bay is included in the Western Hemisphere Shorebird Reserve Network because of its importance as a migratory stopover and overwintering site for tens of thousands of shorebirds. The bay also supports tens of thousands of wintering waterfowl and is the northernmost wintering habitat for large numbers of Long-billed Curlews, Marbled Godwits, and Willets.

KING SALMON King Salmon Avenue is 8 mi north of Loleta, 5 mi south of Eureka, running west off Hwy 101. Drive about 0.5 mi out King Salmon, then park and walk west to the shore. The mouth of the Humboldt Bay is visible from here, a good place to see diving birds as well as Black Brant.

HILFIKER LANE The drive north along the eastern edge of the bay offers vistas of open bay and shoreline to the west that attract waterfowl and raptors. There is a serviceable viewing area of the midbay accessible from Hilfiker Lane at the mouth of the Elk River.

EUREKA The town of Eureka is a sprawling commercial center located at midbay. Birding areas especially recommended for time-limited roaming are King Salmon and Field's Landing, Mad River County Park, Arcata Marsh, and Arcata Bottoms. Arcata Bottoms attracts numerous raptors—harriers, peregrines, and Short-eared Owls—and during high tides, check the fields for waders and gulls.

Humboldt Bay National Wildlife Refuge

WWW.FWS.GOV/HUMBOLDTBAY

The refuge protects the open water of the northern bay for the throngs of waterfowl that gather here, fall through spring. Eelgrass beds submerged within the refuge provide an ecological anchor and account for the importance of the site for Black Brant. The sea goose's reliance on eelgrass forces it to leapfrog up the coast during spring migration from Mexico to the Arctic. Only a few large West Coast estuaries support undisturbed and viable eelgrass beds, and Humboldt Bay provides critical migratory staging habitat for more than 50 percent of the Pacific population.

SALMON CREEK UNIT The refuge has a visitor's center located in the Salmon Creek Unit that has observation decks and the Shorebird Loop Trail (1.7 mi RT). This is a recommend stop to pick up maps and checklists. The Salmon Creek Unit is located about 10 mi south of Eureka. Take the Hookton Road exit from Hwy 101 onto Ranch Road.

HUMBOLDT BAY JETTIES Some of the best places to view waterbirds within the refuge are the North Jetty and the South Spit, which provide access to waterbirds on the Pacific Ocean. Red-necked Grebes, jaegers, Pelagic Cormorants, Black Turnstones, Wandering Tattlers, and Sanderlings are reliable in season.

The Arcata Marsh and Wildlife Sanctuary

569 SOUTH G STREET, ARCATA, CA 95521

(707) 826-2359, (707) 822-LOON (5666) BIRDBOX PHONE FOR RECENT RARE BIRD UPDATES; INFO@ARCATAMARSHFRIENDS.ORG

HTTP://ARCATAMARSHFRIENDS.ORG/CONTACT.PHP

The sanctuary is located 5 mi north of Eureka, just west of Hwy 101 in the town of Arcata. Exit west on Samoa Boulevard, then turn left (south) on G or I Street and proceed 0.5 mi to the sanctuary. You will pass by Allen Gearheart and Hauser marshes to end in a parking lot for Klopp Lake and the Arcata Boat Ramp. The sanctuary includes 307 acres of freshwater marshes, salt marsh with tidal slough, grassy uplands, tidal mudflats, brackish marsh, 5.4 mi of walking and biking paths, and an interpretive center. Bird blinds (where people hide to get close looks at wildlife) and benches are scattered along the trail system, and there are outhouses at Klopp Lake parking lot.

The Arcata Marsh Project is a sterling model of how a city can "undevelop" an environmentally harmful situation—in this case an abandoned landfill—and restore it as an educationally, recreationally, and biologically positive arena. This ecologically sound waste treatment system is managed "to optimize fish and wildlife habitat, public use, recreation, and water quality." The interpretive center has interpretive displays, a nature store, bathrooms, a person to answer relevant questions, and in case it is raining, marsh-view windows. Saturday mornings a free bird walk led by an experienced birder (or two) of the Redwood Region Audubon Society meets in the parking lot at the south end of I Street, and Saturday afternoon a docent-led tour of the marsh and wildlife sanctuary leaves the visitor's center. Check the website or call for times. The morning walk will be strongly oriented to birds and their identification. The later trip may be more general. All levels of experience and all ages are welcome to both events.

While birding for all levels of interest is excellent year-round at Arcata Marsh, the Humboldt Bay area contains many great wildlife areas worth visiting. Checklists and guides to birds and birding in the area are available at the visitor's center.

Northern Humboldt County

TRINIDAD HEAD About 20 mi north of Eureka is the small, picturesque fishing village of Trinidad. Nearby Trinidad Head is a high bluff overlooking Trinidad Bay. A hiking trail around Trinidad Head is slightly strenuous in stretches but affords excellent ocean views to the west. Offshore rocks serve as nesting habitat for seabirds, most noticeably the Common Murre; the Tufted Puffin is rare but possible here. Scan Trinidad Bay for other seabirds, including Marbled Murrelets. Peregrine Falcons often cruise by at eye level, and Black Oystercatchers are reliable on the rocky shore. Humboldt State University operates a marine research station in town that is open to the public. Pelagic birding trips offshore over submarine Trinidad Canyon are often organized by the Redwood Regional Audubon Society.

TRINIDAD STATE BEACH This beach has a trail to Elk Head, another vantage point for viewing seabirds nesting or roosting on rocks offshore. Patrick's Point and Luffenholtz Beach are also worthwhile destinations.

HUMBOLDT LAGOONS STATE PARK (BIG LAGOON AND STONE LAGOON) This park is located 40 mi north of Eureka and 55 mi south of Crescent City on Hwy 101. The main topographic features of interest are Big Lagoon and Stone Lagoon. Big Lagoon is 3.5 mi long, blocked off from the Pacific by a barrier beach several hundred feet wide. The lagoon's marshy habitat is an attractive rest stop for migratory and wintering waterfowl. The mature and magnificent redwood forest east and north of the lagoons supports more nesting Marbled Murrelets than perhaps anywhere else in the state. Birds rare or absent elsewhere along the California Coast—Ruffed Grouse *(Bonasa umbellus)*, Gray Jay *(Perisoreus canadensis)*, and Black-capped Chickadee *(Poecile atricapillus)*—are resident in these forests and north into Prairie Creek Redwoods State Park. Roosevelt Elk *(Cervus canadensis roosevelti)* are often seen in the open fields along Hwy 101 at Stone Lagoon and north to Prairie Creek Redwoods State Park. The park is located on Newton B. Drury Scenic Parkway off Hwy 101, in northeasternmost Humboldt County, 50 mi north of Eureka and 25 mi south of Crescent City.

Del Norte County

Located virtually at the center of the Del Norte County coastline, 20 mi south of the Oregon border, Crescent City is the only incorporated city in the county. Some of the county's most interesting coastal birding sites—Crescent City Harbor, Point St. George, Castle Rock National Wildlife Refuge, Lake Earl, and the inappropriately named Dead Lake—are nearby. The National Audubon Society has designated the 15 mi coastal strip from Point St. George northward to the Smith River as an Important Bird Area (IBA).

CRESCENT CITY HARBOR This harbor provides excellent year-round birding in a discrete and usually quiet area. To access the harbor, take Anchor Way off Hwy 101 just south of the Crescent City limits, drive a short way and park where the harbor is on your right and the ocean is on your left. Many loons, grebes, cormorants, and sea ducks (California's first Steller's Eider was found here) may be seen here, as well as, sometimes, River Otter *(Lontra canadensis)*. Climb the hill at the end of Anchor Way for a good harbor view.

To access the north side of the harbor, continue north on Hwy 101 less than a mile, then bear west on Front Street (following Coastal Access signs), go into the park (yes, there are bathrooms) at the wooden bear with the bow tie, and drive a short way to the water. There is good birding here. You can also drive along the harbor edge west to B Street and left to the over-water parking lot, then walk out the pier for viewing. Among the usual waterbirds, watch for the rare Harlequin and Long-

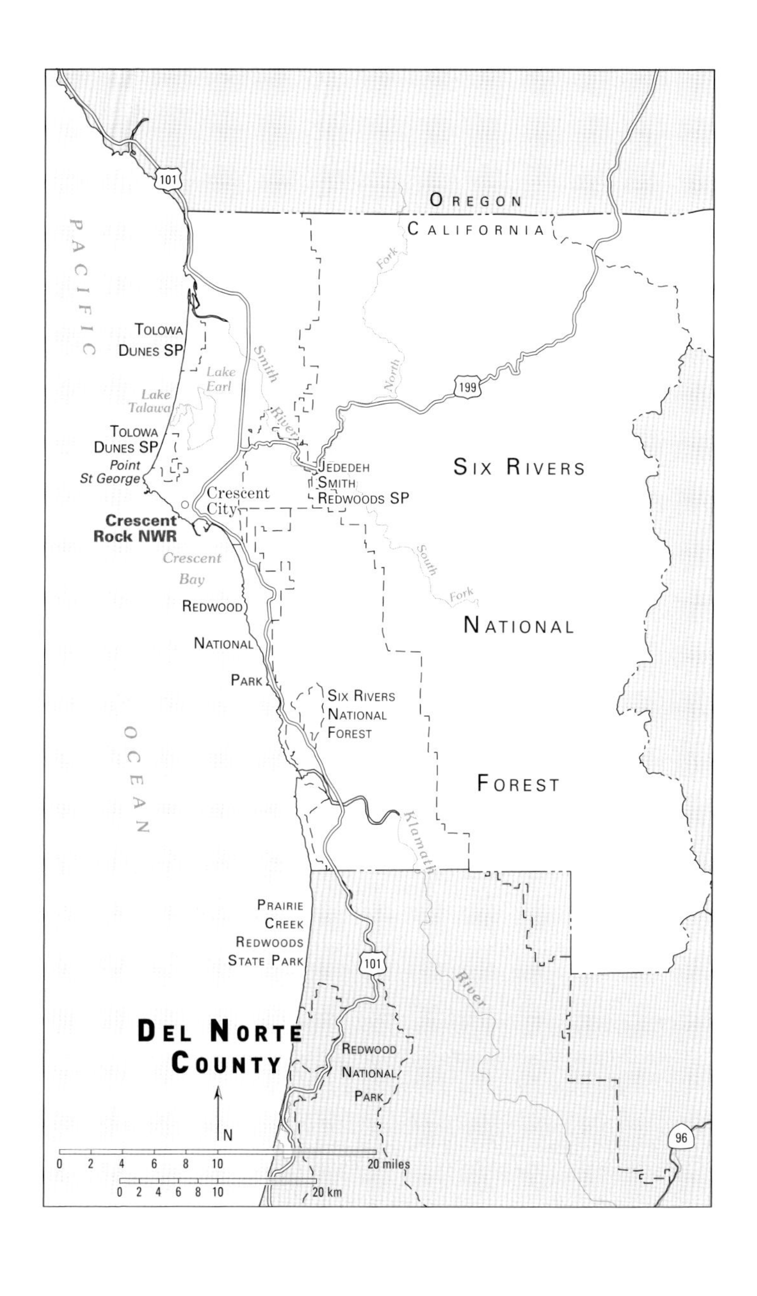

101
Oregon
California
Pacific
Ocean
Tolowa
Dunes SP
Lake
Earl
Lake
Talawa
Smith
River
North
Fork
199
Tolowa
Dunes SP
Point
St George
Jededeh
Smith
Redwoods SP
Six Rivers
National
Forest
Crescent
City
Crescent
Rock NWR
Crescent
Bay
South
Fork
Redwood
National
Park
Six Rivers
National
Forest
Klamath
River
Prairie
Creek
Redwoods
State Park
101
Del Norte
County
Redwood
National
Park
N
96
0 2 4 6 8 10 20 miles
0 2 4 6 8 10 20 km

tailed Ducks. Scan the breakwater for rocky shore birds, most notably Rock Sandpiper *(Calidris ptilocnemis)*, a species that becomes increasingly rare southward.

POINT ST. GEORGE This point, 5 mi NNE of Crescent City Harbor, is a coastal promontory jutting into the Pacific. It is one of the best places in California to see Harlequin Duck and Rock Sandpiper; the duck is fairly reliable, the sandpiper rather rare. Viewing is best from the northwest corner of the bluff. From Crescent City Harbor, take Hwy 1 north to the Northcrest Road, then turn left onto County Road D1/East Washington Blvd, and continue straight 2 miles to North Pebble Beach Road, which will end at Point St. George. There are many islets and sea stacks off Pebble Beach Road, and the biggest one is Castle Rock National Wildlife Refuge.

CASTLE ROCK NATIONAL WILDLIFE REFUGE Covering 14 acres, and 235 ft high at its peak, it is second only to Southeast Farallon Island as the largest seabird colony south of Alaska. Castle Rock and the Farallons are the only two islands on the outer coast of California included in the National Wildlife Refuge System. The rock is over 0.5 mi offshore, so viewing is difficult; a telescope is essential. Castle Rock NWR is closed to the public to prevent disturbance to the seabirds, their habitat, and marine mammals. The primary viewing area for Castle Rock is from Pebble Beach and the mainland at Point St. George. Early morning hours are best. Tens of thousands of Common Murres plus gulls, cormorants, auklets, puffins, and Leach's and Fork-tailed Storm-Petrels nest here. With good binoculars, carpets of the black-and-white murres may be seen during spring, and with a good telescope, you might spot one or two Tufted Puffins sitting at the mouths of their burrows. Noisy colonies of California (and some Steller's) Sea Lions plus Harbor and Elephant Seals crowd the lower cliffs and basal rocks. A Peregrine or two usually have the whole scene under surveillance.

In March and April, tens of thousands of Aleutian Cackling Geese roost on the island at night after feeding in pastures during the day. The dawn fly-off can be astounding. Go left on Washington to the Point St. George Lighthouse parking lot and trails, through coastal bluffs to rocky shores.

TOLOWA DUNES STATE PARK This is a 5,000-plus-acre wildlife area that encompasses Lake Earl and Lake Tolowa, a wetland complex with a 60 mi perimeter, a few miles northwest of

Crescent City. Access to the marshy shoreline and distant open water of Lake Earl is at the end of Old Mill Road. From Hwy 101 take Lake Earl Drive. Go left at Old Mill Road. Proceed 1.4 mi to the wildlife area headquarters, managed by the California Department of Fish and Game. Here are two estuarine lagoons connected by a deep channel and bordered by salt and freshwater marshes. Lake Earl is fresher. Western Grebes nest on the lake perimeters, and the marsh bird and waterfowl numbers in winter can be impressive. A 1 mi loop trail wends through coastal Sitka spruce–grand fir forest. The Lake Earl and Brush and Jordan Creek estuaries may hold flocks of waterfowl or shorebirds in migration. In the marshes look for Northern Harrier, Great Egret, Green Heron, Common Yellowthroat, and Marsh Wren. American Bittern, Virginia Rail, and Sora are some more secretive possibilities.

DEAD LAKE Tolowa Dunes State Park has a 3.5 mi round-trip hike to Dead Lake on the ocean shore, the Dead Lake Trail. At the north end of Crescent City, bear left on Northcrest Drive and follow it 0.5 mi to Old Mill Road. Turn left and drive 1 mi to Sand Hill Road, then turn left and continue a short distance to the road's end at the parking lot. A mile out, you'll come to Dead Lake, a small freshwater lake, a former lumber mill pond. Wood Ducks are common in here. To reach the ocean shore, continue another 0.75 mi along the path.

SMITH RIVER BOTTOMS This is a mosaic of wet pastureland, riparian corridors, and conifer islands, mostly privately owned. There are several ways to explore the area. To access the north side of the river, take Smith River Road west off Hwy 101 and go 0.5 mi to a parking lot. The estuary and mouth are habitat for a variety of year-round resident and migrant birds including Common, Pacific, and Red-throated Loons; Brown Pelican; Western Grebe; Double-crested Cormorant; and a variety of gulls. Large waders—Black-crowned Night Heron, Great Egret, Snowy Egret, and Great Blue Heron—are common here. You may see Harbor Seals basking in the sun, or a family of River Otters. From the south, take Lake Earl Road to Lower Lake Road ending at Pala Road. This route will take you through the pasturelands where thousands of Aleutian Cackling Geese stage during their spring (March and April) migration. In winter Tundra Swans and a variety of raptors are likely here as well. The Smith River hosts one of the few Bank Swallow colonies left in coastal California; in fact, seven species of swallows can be seen here in summer.

SOUTH OF CRESCENT CITY Other suggested bird-searching destinations south of Crescent City are False Klamath Cove (including Lagoon Creek Pond just south of the cove), Endert's Beach cliff trail, and the Klamath River mouth and estuaries (just south of the town of Requa). Sanger Peak (5,862 ft elevation) and vicinity (Rogue River-Siskyou and Six Rivers National Forest) is a remote area of northwest California with access via Knopki Creek Road along gravelly roads not accessible in winter.

OCCURRENCE CHARTS

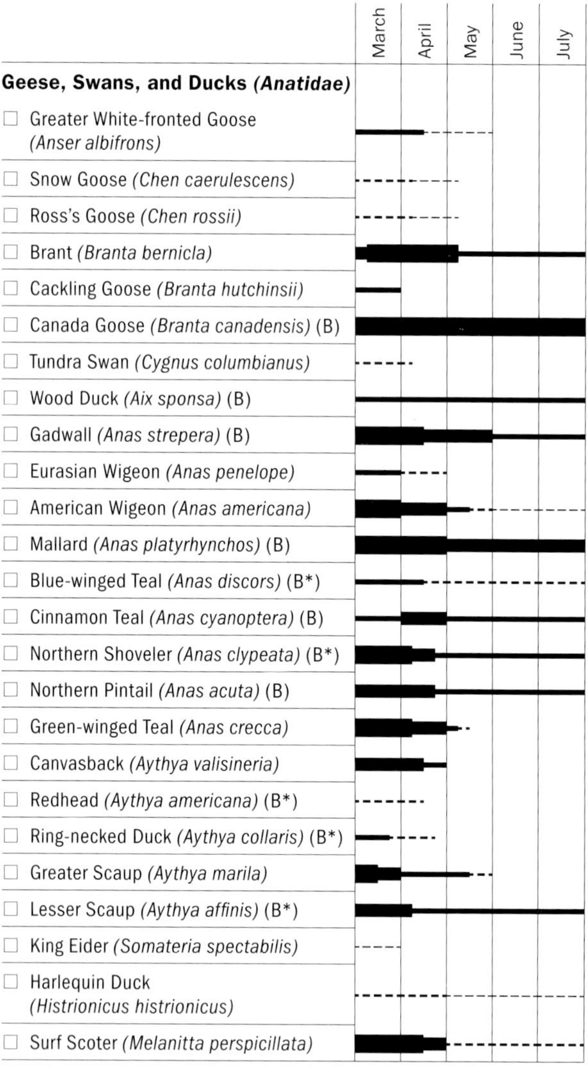

	March	April	May	June	July
Geese, Swans, and Ducks *(Anatidae)*					
☐ Greater White-fronted Goose *(Anser albifrons)*					
☐ Snow Goose *(Chen caerulescens)*					
☐ Ross's Goose *(Chen rossii)*					
☐ Brant *(Branta bernicla)*					
☐ Cackling Goose *(Branta hutchinsii)*					
☐ Canada Goose *(Branta canadensis)* (B)					
☐ Tundra Swan *(Cygnus columbianus)*					
☐ Wood Duck *(Aix sponsa)* (B)					
☐ Gadwall *(Anas strepera)* (B)					
☐ Eurasian Wigeon *(Anas penelope)*					
☐ American Wigeon *(Anas americana)*					
☐ Mallard *(Anas platyrhynchos)* (B)					
☐ Blue-winged Teal *(Anas discors)* (B*)					
☐ Cinnamon Teal *(Anas cyanoptera)* (B)					
☐ Northern Shoveler *(Anas clypeata)* (B*)					
☐ Northern Pintail *(Anas acuta)* (B)					
☐ Green-winged Teal *(Anas crecca)*					
☐ Canvasback *(Aythya valisineria)*					
☐ Redhead *(Aythya americana)* (B*)					
☐ Ring-necked Duck *(Aythya collaris)* (B*)					
☐ Greater Scaup *(Aythya marila)*					
☐ Lesser Scaup *(Aythya affinis)* (B*)					
☐ King Eider *(Somateria spectabilis)*					
☐ Harlequin Duck *(Histrionicus histrionicus)*					
☐ Surf Scoter *(Melanitta perspicillata)*					

Key: (B) = breeds in region; (B*) = breeds very locally or irregularly; (I) = introduced

Aug	Sept	Oct	Nov	Dec	Jan	Feb	Notes

abundant to common · fairly common · rare to uncommon · very rare · extremely rare · irregular

	March	April	May	June	July
☐ White-winged Scoter *(Melanitta fusca)*					
☐ Black Scoter *(Melanitta americana)*					
☐ Long-tailed Duck *(Clangula hyemalis)*					
☐ Bufflehead *(Bucephala albeola)*					
☐ Common Goldeneye *(Bucephala clangula)*					
☐ Barrow's Goldeneye *(Bucephala islandica)*					
☐ Hooded Merganser *(Lophodytes cucullatus)* (B*)					
☐ Common Merganser *(Mergus merganser)* (B)					
☐ Red-breasted Merganser *(Mergus serrator)*					
☐ Ruddy Duck *(Oxyura jamaicensis)* (B*)					
New World Quail *(Odontophoridae)*					
☐ California Quail *(Callipepla californica)* (B)					
Partridges, Grouse, and Turkeys *(Phasianidae)*					
☐ Ring-necked Pheasant *(Phasianus colchicus)* (I) (B)					
☐ Ruffed Grouse *(Bonasa umbellus)* (B)					
☐ Wild Turkey *(Meleagris gallopavo)*(I) (B)					
Loons *(Gaviidae)*					
☐ Red-throated Loon *(Gavia stellata)*					
☐ Pacific Loon *(Gavia pacifica)*					
☐ Common Loon *(Gavia immer)*					
Grebes *(Podicipedidae)*					
☐ Pied-billed Grebe *(Podilymbus podiceps)* (B)					

Key: (B) = breeds in region; (B*) = breeds very locally or irregularly; (I) = introduced

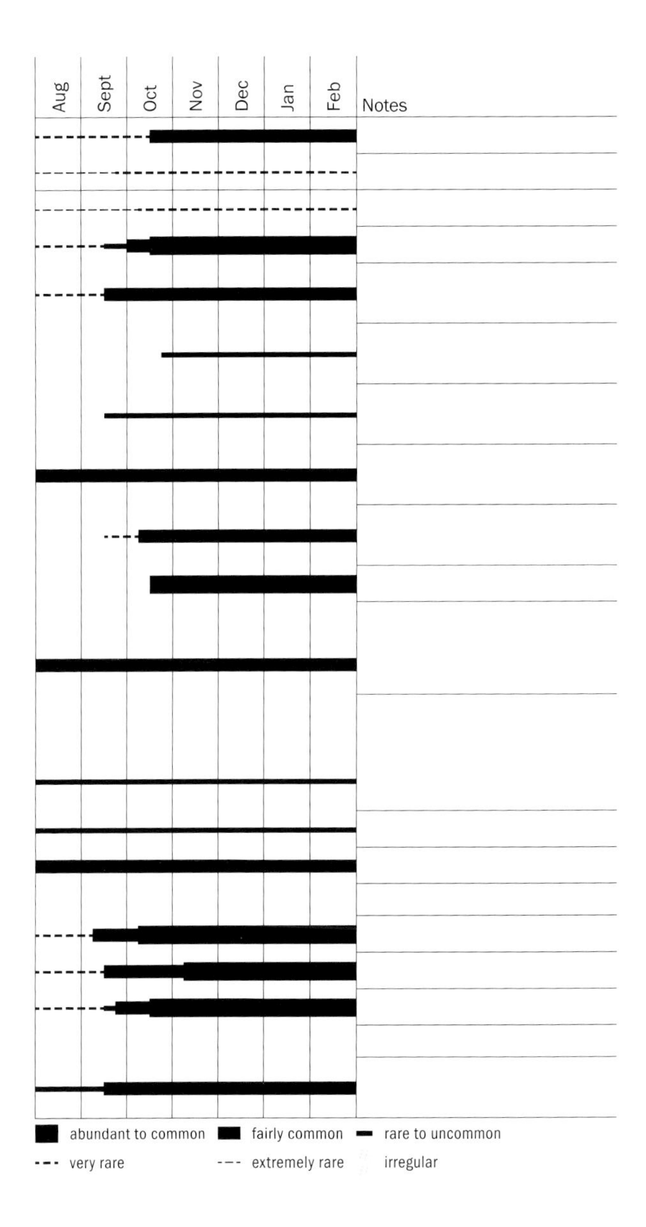

Aug
Sept
Oct
Nov
Dec
Jan
Feb
Notes
abundant to common
fairly common
rare to uncommon
very rare
extremely rare
irregular

	March	April	May	June	July
☐ Horned Grebe *(Podiceps auritus)*					
☐ Red-necked Grebe *(Podiceps grisegena)*					
☐ Eared Grebe *(Podiceps nigricollis)*					
☐ Western Grebe *(Aechmophorus occidentalis)*					
☐ Clark's Grebe *(Aechmophorus clarkii)* (B*)					
Shearwaters and Petrels *(Procellariidae)*					
☐ Northern Fulmar *(Fulmarus glacialis)*					
☐ Sooty Shearwater *(Puffinus griseus)*					
☐ Black-vented Shearwater *(Puffinus opisthomelas)*					
Cormorants *(Phalacrocoracidae)*					
☐ Brandt's Cormorant *(Phalacrocorax penicillatus)* (B)					
☐ Double-crested Cormorant *(Phalacrocorax auritus)* (B)					
☐ Pelagic Cormorant *(Phalacrocorax pelagicus)* (B)					
Pelicans *(Pelecanidae)*					
☐ American White Pelican *(Pelecanus erythrorhynchos)*					
☐ Brown Pelican *(Pelecanus occidentalis)*					
Bitterns, Herons, and Allies *(Ardeidae)*					
☐ American Bittern *(Botaurus lentiginosus)* (B)					
☐ Least Bittern *(Ixobrychus exilis)* (B*)					
☐ Great Blue Heron *(Ardea herodias)* (B)					
☐ Great Egret *(Ardea alba)* (B)					

Key: (B) = breeds in region; (B*) = breeds very locally or irregularly; (I) = introduced

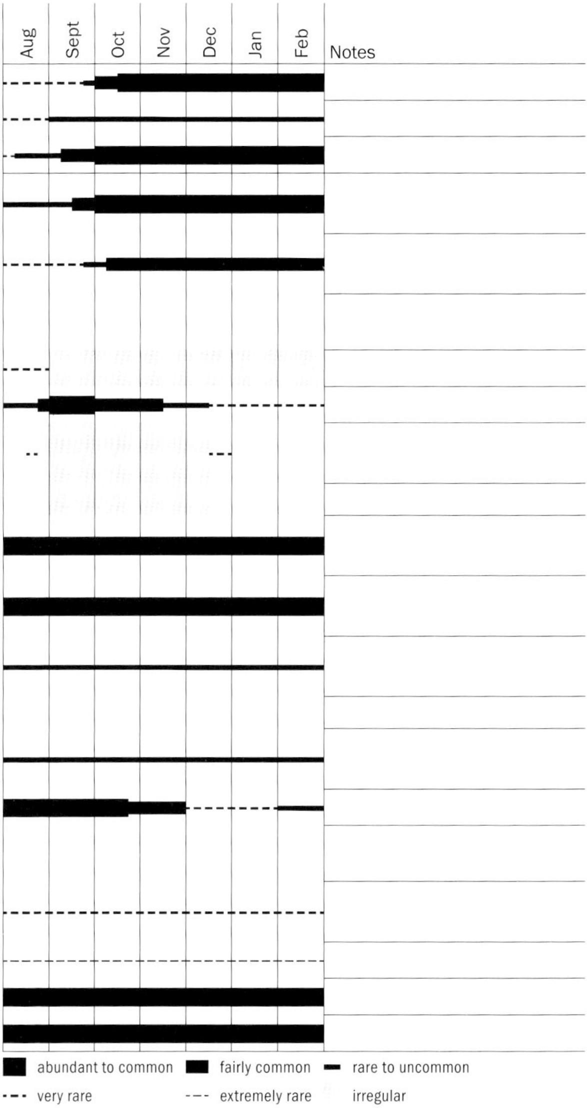

Aug	Sept	Oct	Nov	Dec	Jan	Feb	Notes

abundant to common
fairly common
rare to uncommon
very rare
extremely rare
irregular

	March	April	May	June	July
☐ Snowy Egret *(Egretta thula)* (B)					
☐ Green Heron *(Butorides virescens)* (B)					
☐ Black-crowned Night Heron *(Nycticorax nycticorax)* (B)					
☐ White-faced Ibis *(Plegadis chihi)*					
New World Vultures *(Cathartidae)*					
☐ Turkey Vulture *(Cathartes aura)* (B)					
Ospreys *(Pandionidae)*					
☐ Osprey *(Pandion haliaetus)* (B)					
Kites, Hawks, and Eagles *(Accipitridae)*					
☐ White-tailed Kite *(Elanus leucurus)* (B)					
☐ Bald Eagle *(Haliaeetus leucocephalus)* (B*)					
☐ Northern Harrier *(Circus cyaneus)* (B)					
☐ Sharp-shinned Hawk *(Accipiter striatus)* (B*)					
☐ Cooper's Hawk *(Accipiter cooperii)* (B)					
☐ Red-shouldered Hawk *(Buteo lineatus)* (B)					
☐ Red-tailed Hawk *(Buteo jamaicensis)* (B)					
☐ Ferruginous Hawk *(Buteo regalis)*					
☐ Rough-legged Hawk *(Buteo lagopus)*					
☐ Golden Eagle *(Aquila chrysaetos)* (B)					
Rails, Gallinules, and Coots *(Rallidae)*					
☐ Yellow Rail *(Coturnicops noveboracensis)*					
☐ Black Rail *(Laterallus jamaicensis)* (B*)					
☐ Clapper Rail *(Rallus longirostris)* (B*)					
☐ Virginia Rail *(Rallus limicola)* (B)					

Key: (B) = breeds in region; (B*) = breeds very locally or irregularly; (I) = introduced

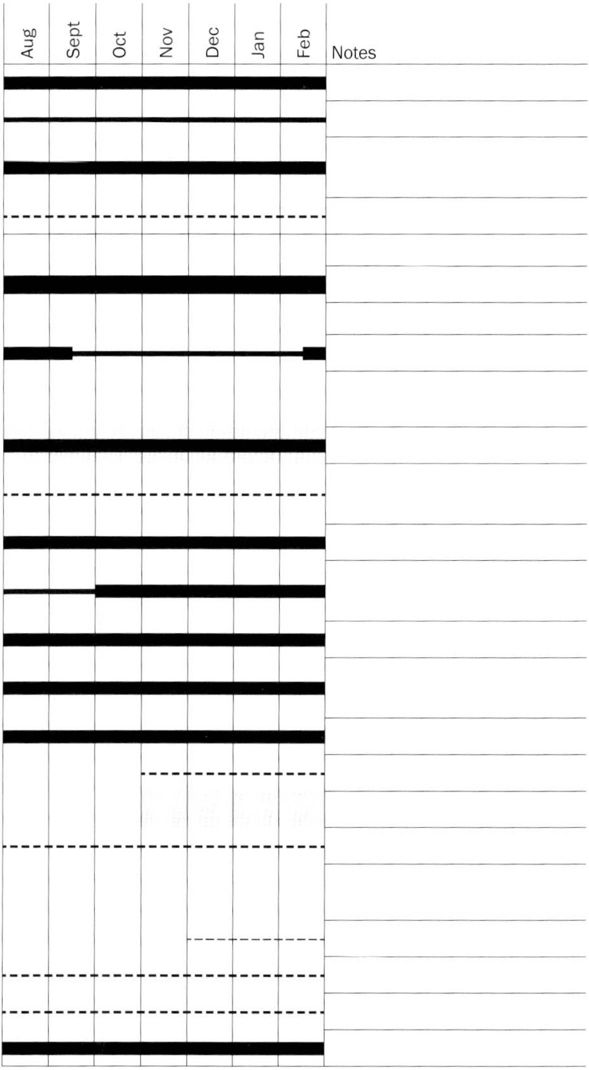

Aug	Sept	Oct	Nov	Dec	Jan	Feb	Notes

abundant to common
fairly common
rare to uncommon
very rare
extremely rare
irregular

	March	April	May	June	July
☐ Sora *(Porzana carolina)* (B)					
☐ Common Gallinule *(Gallinula galeata)* (B)					
☐ American Coot *(Fulica americana)* (B)					
Stilts and Avocets *(Recurviro stridae)*					
☐ Black-necked Stilt *(Himantopus mexicanus)* (B)					
☐ American Avocet *(Recurvirostra americana)* (B)					
Oystercatchers *(Haematopodidae)*					
☐ Black Oystercatcher *(Haematopus bachmani)* (B)					
Plovers and Lapwings *(Charadriidae)*					
☐ Black-bellied Plover *(Pluvialis squatarola)*					
☐ American Golden-Plover *(Pluvialis dominica)*					
☐ Pacific Golden-Plover *(Pluvialis fulva)*					
☐ Snowy Plover *(Charadrius nivosus)* (B*)					
☐ Semipalmated Plover *(Charadrius semipalmatus)*					
☐ Killdeer *(Charadrius vociferus)* (B)					
Sandpipers, Phalaropes, and Allies *(Scolopacidae)*					
☐ Spotted Sandpiper *(Actitis macularius)* (B*)					
☐ Wandering Tattler *(Tringa incana)*					
☐ Greater Yellowlegs *(Tringa melanoleuca)*					
☐ Willet *(Tringa semipalmata)*					
☐ Lesser Yellowlegs *(Tringa flavipes)*					
☐ Whimbrel *(Numenius phaeopus)*					

Key: (B) = breeds in region; (B*) = breeds very locally or irregularly; (I) = introduced

Aug	Sept	Oct	Nov	Dec	Jan	Feb	Notes

abundant to common — fairly common — rare to uncommon — very rare — extremely rare — irregular

	March	April	May	June	July
☐ Long-billed Curlew *(Numenius americanus)*					
☐ Marbled Godwit *(Limosa fedoa)*					
☐ Ruddy Turnstone *(Arenaria interpres)*					
☐ Black Turnstone *(Arenaria melanocephala)*					
☐ Red Knot *(Calidris canutus)*					
☐ Surfbird *(Calidris virgata)*					
☐ Ruff *(Calidris pugnax)*					
☐ Sharp-tailed Sandpiper *(Calidris acuminata)*					
☐ Stilt Sandpiper *(Calidris himantopus)*					
☐ Sanderling *(Calidris alba)*					
☐ Dunlin *(Calidris alpina)*					
☐ Rock Sandpiper *(Calidris ptilocnemis)*					
☐ Baird's Sandpiper *(Calidris bairdii)*					
☐ Least Sandpiper *(Calidris minutilla)*					
☐ Pectoral Sandpiper *(Calidris melanotos)*					
☐ Semipalmated Sandpiper *(Calidris pusilla)*					
☐ Western Sandpiper *(Calidris mauri)*					
☐ Short-billed Dowitcher *(Limnodromus griseus)*					
☐ Long-billed Dowitcher *(Limnodromus scolopaceus)*					
☐ Wilson's Snipe *(Gallinago delicata)*					
☐ Wilson's Phalarope *(Phalaropus tricolor)*					
☐ Red-necked Phalarope *(Phalaropus lobatus)*					
☐ Red Phalarope *(Phalaropus fulicarius)*					

Key: (B) = breeds in region; (B*) = breeds very locally or irregularly; (I) = introduced

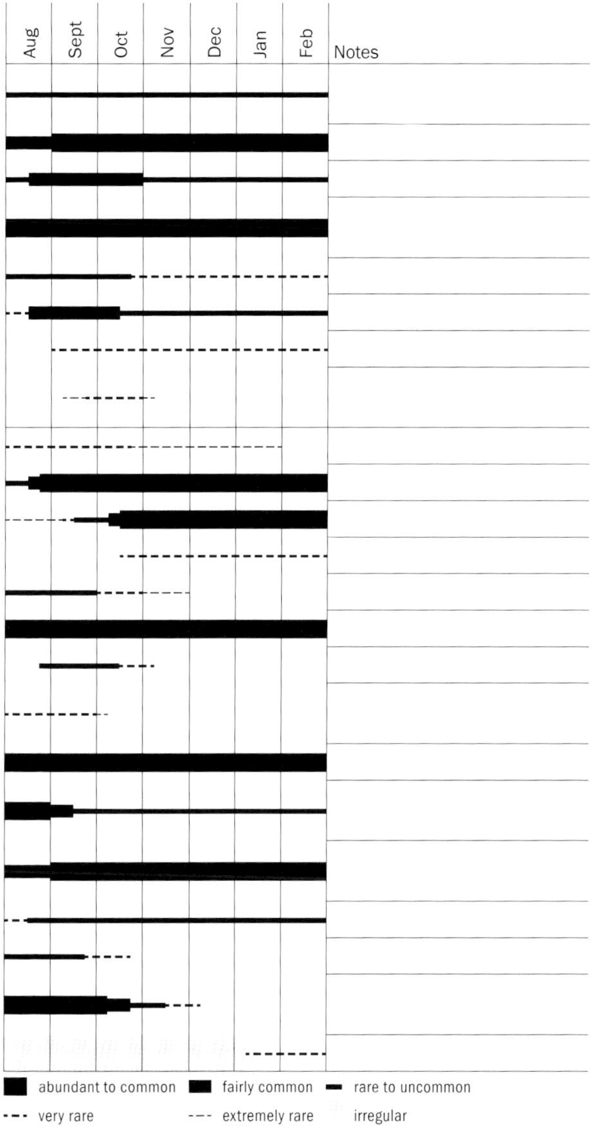

Aug	Sept	Oct	Nov	Dec	Jan	Feb	Notes

abundant to common
fairly common
rare to uncommon
very rare
extremely rare
irregular

	March	April	May	June	July
Skuas *(Stercorariidae)*					
☐ Parasitic Jaeger *(Stercorarius parasiticus)*					
Auks, Murres, and Puffins *(Alcidae)*					
☐ Common Murre *(Uria aalge)* (B)					
☐ Pigeon Guillemot *(Cepphus columba)* (B)					
☐ Marbled Murrelet *(Brachyramphus marmoratus)* (B*)					
☐ Ancient Murrelet *(Synthliboramphus antiquus)*					
☐ Cassin's Auklet *(Ptychoramphus aleuticus)* (B)					
☐ Rhinoceros Auklet *(Cerorhinca monocerata)* (B)					
☐ Tufted Puffin *(Fratercula cirrhata)* (B)					
Gulls, Terns, and Skimmers *(Laridae)*					
☐ Black-legged Kittiwake *(Rissa tridactyla)*					
☐ Bonaparte's Gull *(Chroicocephalus philadelphia)*					
☐ Laughing Gull *(Leucophaeus atricilla)*					
☐ Franklin's Gull *(Leucophaeus pipixcan)*					
☐ Heermann's Gull *(Larus heermanni)*					
☐ Mew Gull *(Larus canus)*					
☐ Ring-billed Gull *(Larus delawarensis)*					
☐ Western Gull *(Larus occidentalis)* (B)					
☐ California Gull *(Larus californicus)* (B)					
☐ Herring Gull *(Larus argentatus)*					
☐ Thayer's Gull *(Larus thayeri)*					
☐ Glaucous-winged Gull *(Larus glaucescens)*					

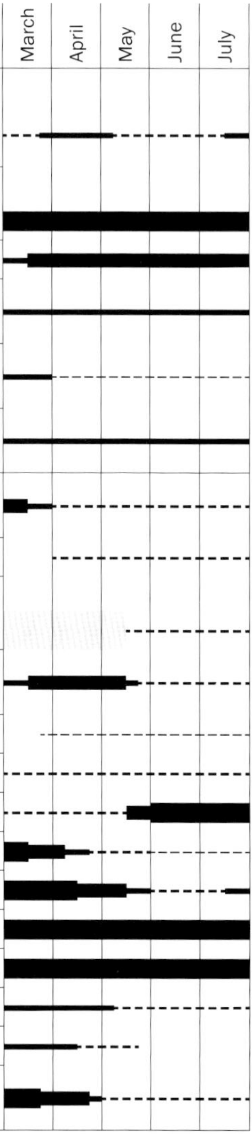

Key: (B) = breeds in region; (B*) = breeds very locally or irregularly; (I) = introduced

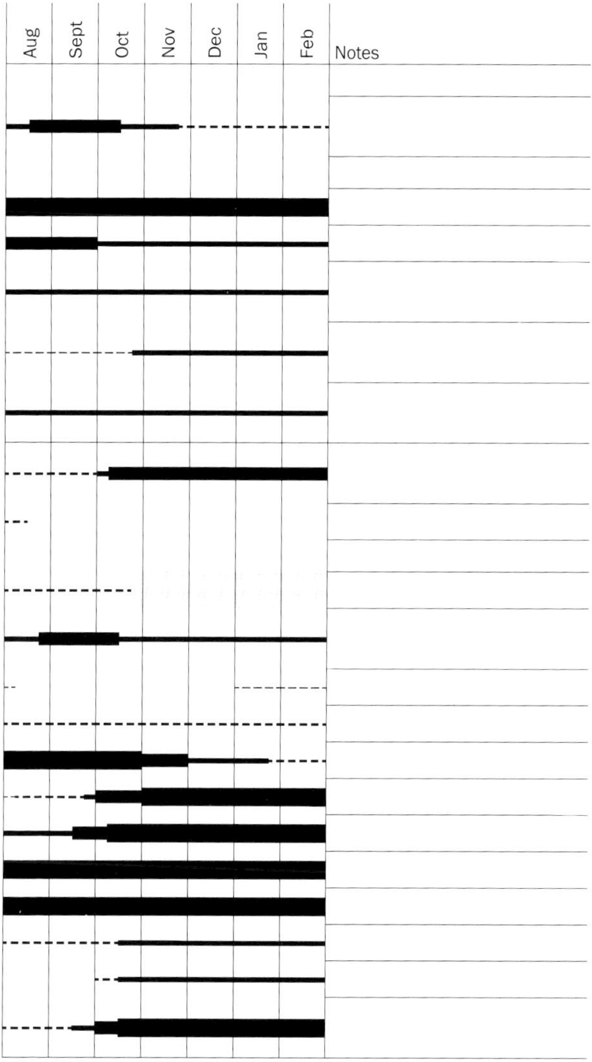

Aug	Sept	Oct	Nov	Dec	Jan	Feb	Notes

abundant to common
fairly common
rare to uncommon
very rare
extremely rare
irregular

	March	April	May	June	July
☐ Glaucous Gull *(Larus hyperboreus)*					
☐ Least Tern *(Sternula antillarum)* (B*)					
☐ Caspian Tern *(Hydroprogne caspia)* (B)					
☐ Black Tern *(Chlidonias niger)*					
☐ Common Tern *(Sterna hirundo)*					
☐ Arctic Tern *(Sterna paradisaea)*					
☐ Forster's Tern *(Sterna forsteri)* (B)					
☐ Elegant Tern *(Thalasseus elegans)*					
☐ Black Skimmer *(Rynchops niger)* (B*)					
Pigeons and Doves *(Columbidae)*					
☐ Rock Pigeon *(Columba livia)* (I)					
☐ Band-tailed Pigeon *(Patagioenas fasciata)* (B)					
☐ Eurasian Collared-Dove *(Streptopelia decaocto)* (I) (B)					
☐ Mourning Dove *(Zenaida macroura)* (B)					
Barn Owls *(Tytonidae)*					
☐ Barn Owl *(Tyto alba)* (B)					
Typical Owls *(Strigidae)*					
☐ Western Screech Owl *(Megascops kennicottii)* (B)					
☐ Great Horned Owl *(Bubo virginianus)* (B)					
☐ Northern Pygmy-Owl *(Glaucidium gnoma)* (B)					
☐ Burrowing Owl *(Athene cunicularia)* (very local) (B*)					
☐ Spotted Owl *(Strix occidentalis)* (local) (B)					
☐ Barred Owl *(Strix varia)* (local) (B)					
☐ Northern Saw-whet Owl *(Aegolius acadicus)* (B)					

Key: (B) = breeds in region; (B*) = breeds very locally or irregularly; (I) = introduced

Aug	Sept	Oct	Nov	Dec	Jan	Feb	Notes

abundant to common — fairly common — rare to uncommon

very rare — extremely rare — irregular

	March	April	May	June	July
Goatsuckers *(Caprimulgidae)*					
☐ Common Nighthawk *(Chordeiles minor)* (B*)					
☐ Common Poorwill *(Phalaenoptilus nuttallii)* (local) (B*)					
Swifts *(Apodidae)*					
☐ Black Swift *(Cypseloides niger)* (local) (B*)					
☐ Vaux's Swift *(Chaetura vauxi)* (B*)					
☐ White-throated Swift *(Aeronautes saxatalis)* (local) (B*)					
Hummingbirds *(Trochilidae)*					
☐ Anna's Hummingbird *(Calypte anna)* (B)					
☐ Rufous Hummingbird *(Selasphorus rufus)*					
☐ Allen's Hummingbird *(Selasphorus sasin)* (B)					
Kingfishers *(Alcedinidae)*					
☐ Belted Kingfisher *(Megaceryle alcyon)* (B)					
Woodpeckers and Allies *(Picidae)*					
☐ Acorn Woodpecker *(Melanerpes formicivorus)* (B)					
☐ Red-breasted Sapsucker *(Sphyrapicus ruber)* (B)					
☐ Nuttall's Woodpecker *(Picoides nuttallii)* (B)					
☐ Downy Woodpecker *(Picoides pubescens)* (B)					
☐ Hairy Woodpecker *(Picoides villosus)* (B)					
☐ Northern Flicker *(Colaptes auratus)* (B)					
☐ Pileated Woodpecker *(Dryocopus pileatus)* (B)					

Key: (B) = breeds in region; (B*) = breeds very locally or irregularly; (I) = introduced

Aug	Sept	Oct	Nov	Dec	Jan	Feb	Notes

abundant to common | fairly common | rare to uncommon
very rare | extremely rare | irregular

	March	April	May	June	July
Caracaras and Falcons *(Falconidae)*					
☐ American Kestrel *(Falco sparverius)* (B)					
☐ Merlin *(Falco columbarius)*					
☐ Peregrine Falcon *(Falco peregrinus)* (B)					
☐ Prairie Falcon *(Falco mexicanus)*					
Tyrant Flycantchers *(Tyrannidae)*					
☐ Olive-sided Flycatcher *(Contopus cooperi)* (B)					
☐ Western Wood-Pewee *(Contopus sordidulus)* (B)					
☐ Willow Flycatcher *(Empidonax traillii)*					
☐ Pacific-slope Flycatcher *(Empidonax difficilis)* (B)					
☐ Black Phoebe *(Sayornis nigricans)* (B)					
☐ Say's Phoebe *(Sayornis saya)*					
☐ Ash-throated Flycatcher *(Myiarchus cinerascens)* (B)					
☐ Western Flycatcher *(Tyrannus vertivalis)*					
Shrikes *(Laniidae)*					
☐ Loggerhead Shrike *(Lanius ludovicianus)*					
Vireos *(Vireonidae)*					
☐ Cassin's Vireo *(Vireo cassinii)* (B)					
☐ Hutton's Vireo *(Vireo huttoni)* (B)					
☐ Warbling Vireo *(Vireo gilvus)* (B)					
Crows and Jays *(Corvidae)*					
☐ Steller's Jay *(Cyanocitta stelleri)* (B)					
☐ Western Scrub-Jay *(Aphelocoma californica)* (B)					
☐ American Crow *(Corvus brachyrhynchos)* (B)					
☐ Common Raven *(Corvus corax)* (B)					

Key: (B) = breeds in region; (B*) = breeds very locally or irregularly; (I) = introduced

Aug	Sept	Oct	Nov	Dec	Jan	Feb	Notes

abundant to common · fairly common · rare to uncommon · very rare · extremely rare · irregular

	March	April	May	June	July
Larks *(Alaudidae)*					
☐ Horned Lark *(Eremophila alpestris)* (local) (B)					
Swallows *(Hirundinidae)*					
☐ Purple Martin *(Progne subis)* (B)					
☐ Tree Swallow *(Tachycineta bicolor)* (B)					
☐ Violet-green Swallow *(Tachycineta thalassina)* (B)					
☐ Northern Rough-winged Swallow *(Stelgidopteryx serripennis)* (B)					
☐ Bank Swallow *(Riparia riparia)* (local) (B)					
☐ Cliff Swallow *(Petrochelidon pyrrhonota)* (B)					
☐ Barn Swallow *(Hirundo rustica)* (B)					
Chickadees and Titmice *(Paridae)*					
☐ Black-capped Chickadee *(Poecile atricapillus)* (local-north) (B)					
☐ Chestnut-backed Chickadee *(Poecile rufescens)* (B)					
☐ Oak Titmouse *(Baeolophus inornatus)* (B)					
Long-tailed Tits and Bushtits *(Aegithalidae)*					
☐ Bushtit *(Psaltriparus minimus)* (B)					
Nuthatches *(Sittidae)*					
☐ Red-breasted Nuthatch *(Sitta canadensis)* (B)					
☐ White-breasted Nuthatch *(Sitta carolinensis)* (B)					
☐ Pygmy Nuthatch *(Sitta pygmaea)* (Mendocino Co. south) (B)					
Creepers *(Certhiidae)*					
☐ Brown Creeper *(Certhia americana)* (B)					

Key: (B) = breeds in region; (B*) = breeds very locally or irregularly; (I) = introduced

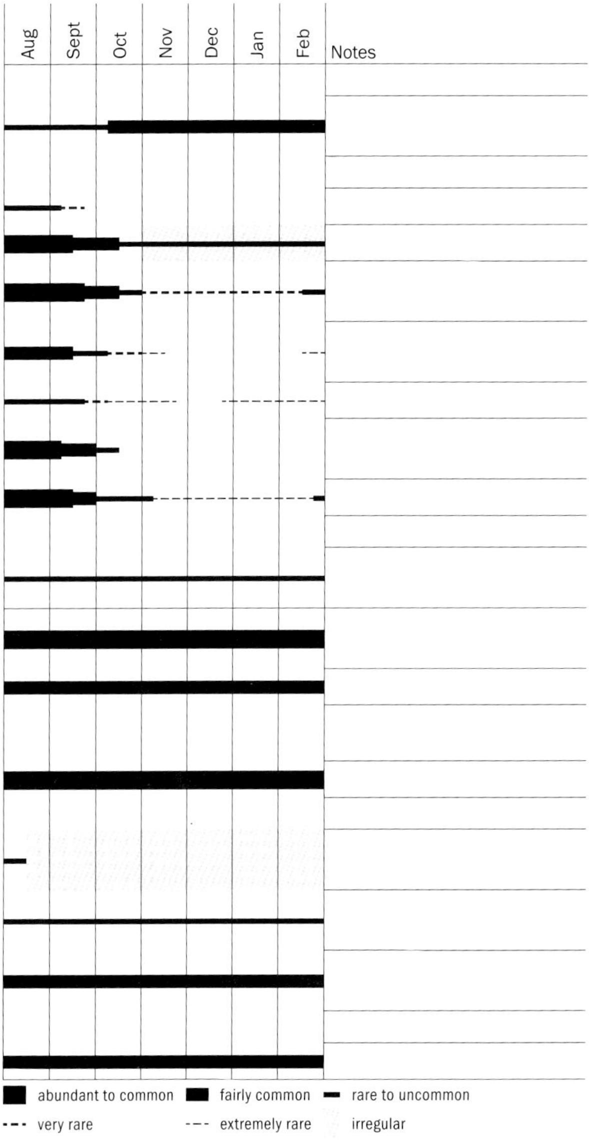

Aug	Sept	Oct	Nov	Dec	Jan	Feb	Notes

abundant to common
fairly common
rare to uncommon
very rare
extremely rare
irregular

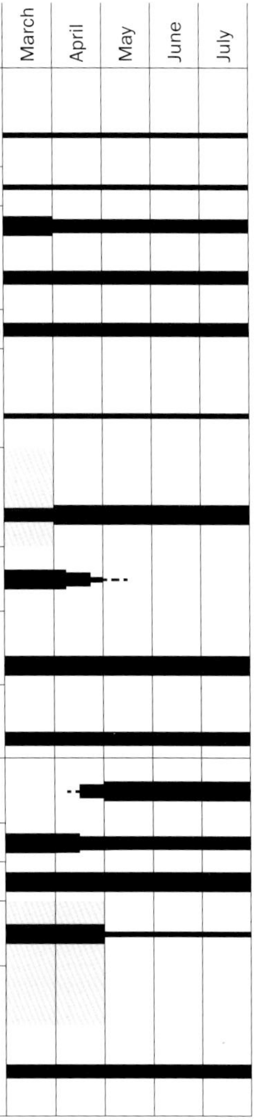

	March	April	May	June	July
Wrens *(Troglodytidae)*					
☐ Rock Wren *(Salpinctes obsoletus)* (local) (B)					
☐ House Wren *(Troglodytes aedon)* (B)					
☐ Pacific Wren *(Troglodytes pacificus)* (B)					
☐ Marsh Wren *(Cistothorus palustris)* (local) (B)					
☐ Bewick's Wren *(Thryomanes bewickii)* (B)					
Dippers (*Cinclidae*)					
☐ American Dipper *(Cinclus mexicanus)* (local) (B*)					
Kinglets (*Regulidae*)					
☐ Golden-crowned Kinglet *(Regulus satrapa)* (B)					
☐ Ruby-crowned Kinglet *(Regulus calendula)* (B)					
Sylvid Warblers *(Sylviidae)*					
☐ Wrentit *(Chamaea fasciata)* (B)					
Thrushes *(Turdidae)*					
☐ Western Bluebird *(Sialia mexicana)* (B)					
☐ Swainson's Thrush *(Catharus ustulatus)* (B)					
☐ Hermit Thrush *(Catharus guttatus)* (B)					
☐ American Robin *(Turdus migratorius)* (B)					
☐ Varied Thrush *(Ixoreus naevius)* (nests in N. counties) (B)					
Mockingbirds and Thrashers *(Mimidae)*					
☐ California Thrasher *(Toxostoma redivivum)* (local in south counties) (B*)					

Key: (B) = breeds in region; (B*) = breeds very locally or irregularly; (I) = introduced

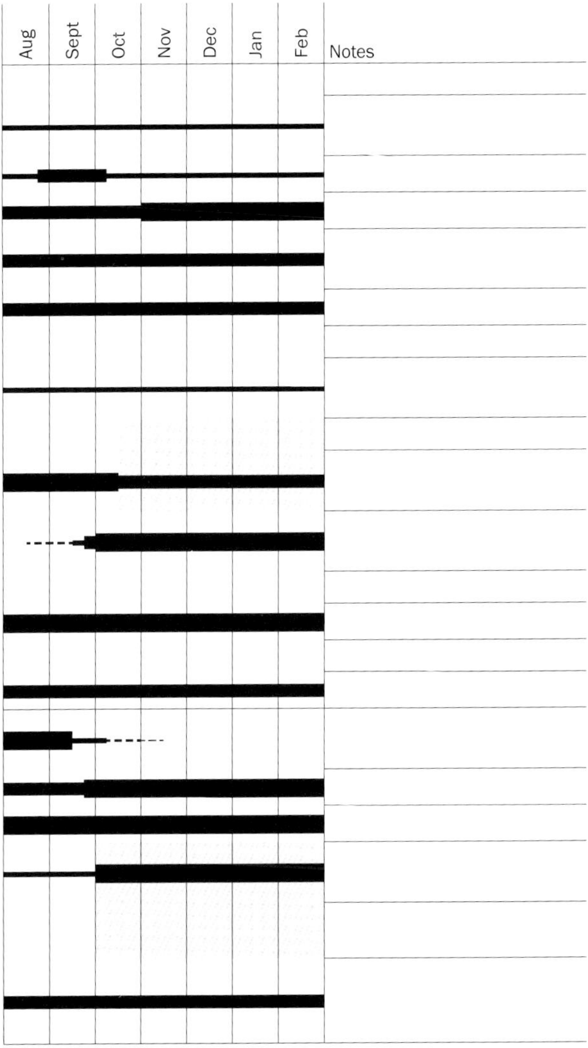

Aug	Sept	Oct	Nov	Dec	Jan	Feb	Notes

abundant to common
fairly common
rare to uncommon
very rare
extremely rare
irregular

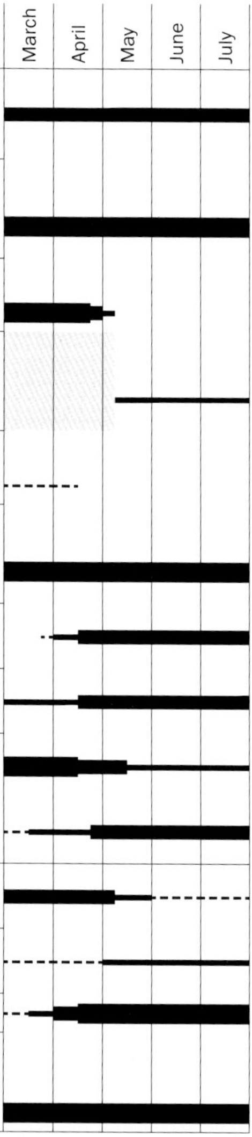

	March	April	May	June	July
☐ Northern Mockingbird *(Mimus polyglottos)* (more common in south) (B)					
Starlings *(Sturnidae)*					
☐ European Starling *(Sturnus vulgaris)* (I) (B)					
Wagtails and Pipits *(Motacillidae)*					
☐ American Pipit *(Anthus rubescens)* (B)					
Waxwings *(Bombycillidae)*					
☐ Cedar Waxwing *(Bombycilla cedrorum)* (B)					
Longspurs *(Calcariidae)*					
☐ Lapland Longspur *(Calcarius lapponicus)*					
Wood-Warblers (Parulidae)					
☐ Orange-crowned Warbler *(Oreothlypis celata)* (B)					
☐ MacGillivray's Warbler *(Geothlypis tolmiei)* (B)					
☐ Common Yellowthroat *(Geothlypis trichas)* (B)					
☐ Yellow-rumped Warbler *(Setophaga coronata)* (B)					
☐ Black-throated Gray Warbler *(Setophaga nigrescens)* (B)					
☐ Townsend's Warbler *(Setophaga townsendi)*					
☐ Hermit Warbler *(Setophaga occidentalis)* (B)					
☐ Wilson's Warbler *(Cardellina pusilla)* (B)					
Towhees, Sparrows, and Allies *(Emberizidae)*					
☐ Spotted Towhee *(Pipilo maculatus)* (B)					

Key: (B) = breeds in region; (B*) = breeds very locally or irregularly; (I) = introduced

Aug	Sept	Oct	Nov	Dec	Jan	Feb	Notes

abundant to common · fairly common · rare to uncommon · very rare · extremely rare · irregular

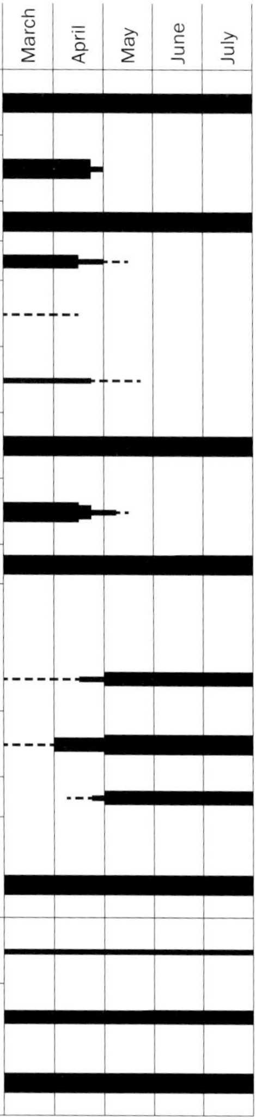

	March	April	May	June	July
☐ California Towhee *(Melozone crissalis)* (more common to south) (B)					
☐ Fox Sparrow *(Passerella iliaca)* (several distinct races)					
☐ Song Sparrow *(Melospiza melodia)* (B)					
☐ Lincoln's Sparrow *(Melospiza lincolnii)*					
☐ Swamp Sparrow *(Melospiza georgiana)* (local)					
☐ White-throated Sparrow *(Zonotrichia albicollis)*					
☐ White-crowned Sparrow *(Zonotrichia leucophrys)* (B)					
☐ Golden-crowned Sparrow *(Zonotrichia atricapilla)*					
☐ Dark-eyed Junco *(Junco hyemalis)* (B)					
Tanagers, Grosbeaks, and Buntings *(Cardinalidae)*					
☐ Western Tanager *(Piranga ludoviciana)* (B)					
☐ Black-headed Grosbeak *(Pheucticus melanocephalus)* (B)					
☐ Lazuli Bunting *(Passerina amoena)* (B)					
Blackbirds *(Icteridae)*					
☐ Red-winged Blackbird *(Agelaius phoeniceus)* (B)					
☐ Tricolored Blackbird *(Agelaius tricolor)* (B)					
☐ Western Meadowlark *(Sturnella neglecta)* (B)					
☐ Brewer's Blackbird *(Euphagus cyanocephalus)* (B)					

Key: (B) = breeds in region; (B*) = breeds very locally or irregularly; (I) = introduced

Aug	Sept	Oct	Nov	Dec	Jan	Feb	Notes

abundant to common — fairly common — rare to uncommon

very rare — extremely rare — irregular

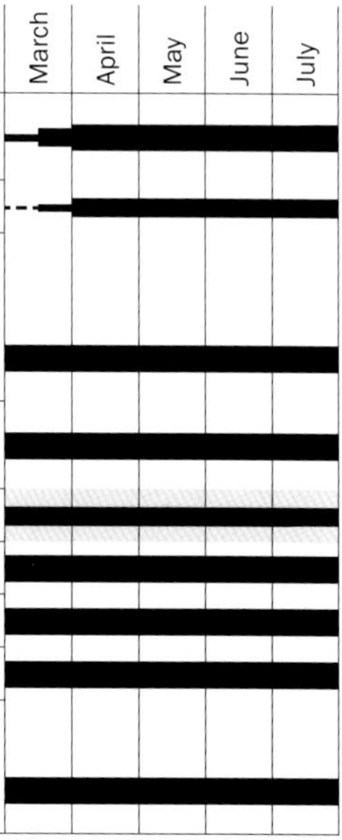

	March	April	May	June	July
☐ Brown-headed Cowbird *(Molothrus ater)* (B)					
☐ Bullock's Oriole *(Icterus bullockii)* (B)					
Fringilline and Cardueline Finches and Allies *(Fringillidae)*					
☐ House Finch *(Haemorhous mexicanus)* (B)					
☐ Purple Finch *(Haemorhous purpureus)* (B)					
☐ Red Crossbill *(Loxia curvirostra)* (B)					
☐ Pine Siskin *(Spinus pinus)* (B)					
☐ Lesser Goldfinch *(Spinus psaltria)* (B)					
☐ American Goldfinch *(Spinus tristis)* (B)					
Old World Sparrows *(Passeridae)*					
☐ House Sparrow *(Passer domesticus)* (I) (B)					

Key: (B) = breeds in region; (B*) = breeds very locally or irregularly; (I) = introduced

Aug	Sept	Oct	Nov	Dec	Jan	Feb	Notes

abundant to common — fairly common — rare to uncommon

very rare — extremely rare — irregular

GLOSSARY

Barbs, barbules, and barbicels The parts of feathers. Barbs branch off from the central shaft (rachis) of contour feathers; barbules, in turn, branch off the barbs. Barbicels are tiny hooks at the ends of the barbules that interlock with the adjacent barbs and barbules to give the feather structural integrity.

Benthic On the bottom under water.

Biological species concept "Species are groups of interbreeding natural populations that are reproductively isolated from other such groups" (Mayr 1942).

Brackish Refers to a mixture of ocean water and freshwater; also refers to marsh vegetation at the edge, as brackish marsh.

Cattails Tall reeds, in the genus *Typha,* found in fresh and brackish marshes.

Clade A group consisting of descendants from a single ancestor (from a Greek root, meaning "branch").

Coastal scrub A widespread habitat of dense but low-growing brush dominated by coyote bush, bush lupine, blackberries, and poison oak. Also called maritime chaparral.

Congener A closely related species, literally one in the same genus.

Cooperative breeder A species that uses a relatively rare nesting strategy that permits more than two birds of the same species to help rear chicks from one nest (e.g., Acorn Woodpecker, Pygmy Nuthatch).

Crepuscular Active in the twilight hours of dawn and dusk.

Cryptic coloration Plumage pattern that blends well with the habitat and is therefore difficult to see.

Culmen The upper edge of the upper bill (maxilla).

Dabbling duck Waterfowl (almost all in the genus *Anas*) that does not dive, but feeds on the surface of the water, "tipping up" or skimming the water's surface with its bill to feed on aquatic vegetation.

Decurved Curved downward, in reference to a bird's bill shape. (See recurved.)

Dihedral Refers to the angle at which the wings are inclined upward (above the body plane) while soaring.

Dimorphism Literally "having two forms," used to describe differences in morphology within a single species.

Diving duck Waterfowl that normally dives completely underwater to search for food.

Dorsal Refers to the upper side of a bird's body; opposite of ventral.

Down Soft, fluffy feathers that lack barbules or structure that allows flight.

Endemic Refers to a population of animals that is native to, or restricted to, a particular geographic locality.

Estero A Spanish synonym for *estuary*.

Extirpated Refers to a population that has become locally extinct.

Formative plumage "Any plumage present only in the first year of life and lacking a counterpart in subsequent molt cycles" (Howell 2010).

GISS (Also "jiss" or "jizz.") An acronym for "general impression of shape and structure," used to describe the ineffable sense of a species.

Glaucous An old English allusion to *bluish*.

Gorget Throat and neck area of a bird's plumage; generally refers to hummingbirds.

Gular pouch The featherless and rather elastic skin of a bird's throat (e.g., in pelicans and cormorants) that expands when fish are swallowed.

Hallux The hind toe of a bird.

Holarctic Refers to the northern continents of the Earth, which can be divided into two subregions: the palearctic (Northern Africa and Eurasia) and the neartic (North America and Greenland); used in reference to animal distribution, as in "the Caspian Tern is a holarctic species." Synonymous with circumboreal.

Hurling Regurgitating pellets of undigested portions of prey (hair, bones) by predators.

Hybrid Offspring of two genetically differing parents, usually of different species.

Hypercoastal Refers to distribution restricted to the immediate coast (e.g., that of the Black Oystercatcher).

Intergrade A bird that appears to be intermediate between two subspecies.

Iridescence Shimmering or lustrous colors that change as the light refracts off the feathers, as do a male hummingbird's gorget or a male Mallard's head.

Irruptive Refers to a sudden or cyclical population increase in response to ecological conditions.

Lamella One of a series of ridges on the inner edge of the bill that allow a bird to siphon food from the water or mud, or to strip food from aquatic vegetation. Plural is lamellae.

Leucistic Refers to paleness of coloration. Another term for hypomelanistic.

Lore The triangular patch of feathering between the eye and the bill; the loreal region.

Lumped Refers to the merger of two species into one by taxonomists (e.g., Bullock's and Baltimore Orioles were "lumped" into one species, the Northern Oriole).

Mantle The back, or dorsal, feathering located between the wings, often used in reference to the back color of gulls.

Microtus The genus name, and the common term, for the meadow vole, an important prey item of some raptors (e.g., White-tailed Kite).

Migrant trap Isolated habitat island (such as a clump of conifers on a coastal bluff) that attracts migrant land birds for rest and food opportunities.

Molt (Also moult.) A cyclical process of feather growth and restoration that occurs once or twice a year in most species, usually after nesting or prior to migration, although there is tremendous variation in the timing and sequencing.

Monophyletic Derived from the same ancestral taxon.

Monotypic One of a kind, as in "Osprey is in a monotypic genus."

Morph A distinctive plumage coloration retained through the life of the individual.

Nearctic The biogeographic region that includes the Arctic and the temperate areas of North America and Greenland. See holarctic.

Nocturnal Active at night.

Omnivorous Feeding on a diet that includes plant and animal matter, maybe even carrion.

Orbital ring A narrow, unfeathered ring of flesh around the eye. The color of the orbital ring is variable among most gull species, an aid to identification.

Palearctic The biogeographic region that includes northern Africa and Eurasia. See holarctic.

Pallid Pale.

Pampas Extensive grassland biome of South America with a windy, warm, and humid climate.

Patagium The leading edge of the wing, between the "wrist" and the body (as in "patagial patch").

Peeps Name applied to several species of the smallest flocking sandpipers (genus *Calidris*) often difficult to identify at species level.

Pelagic Associated with the open ocean.

Phylogenetic species concept "A species is the smallest diagnosable cluster of individual organisms within which there is a parental pattern of ancestry and descent" (Cracraft 1983).

Plunge divers Birds that dive head first into the water when foraging (e.g., Brown Pelicans, most terns, kingfishers).

Poachers Humans who kill wildlife illegally—hunting game species out of season, hunting anything in parks and sanctuaries, or hunting protected species anywhere at any time.

Polygamous Using polygamy as a breeding strategy; *polygynous* refers to polygyny; *polyandrous* refers to polyandry.

Polygamy A breeding strategy in which either the male or the female of a species mates with more than one individual of the opposite gender. In polygyny males mate with more than one female; in polyandry females mate with more than one male.

Polymorphic Having two or more plumage patterns within one species (e.g., Northern Fulmar, Red-tailed Hawk).

Predator An animal that kills other animals for food. Hawks, owls, shrikes, and egrets, among others, are predators.

Primaries Outer flight feathers of a bird's wing.

Raft A tightly packed flock of swimming or "loafing" birds on the water's surface.

Rectrices The tail feathers of a bird.

Recurved Curved upward, in reference to a bird's bill shape (see decurved). Avocets have recurved bills and belong to the family Recurvirostridae, as do stilts.

Remiges The major wing feathers of a bird—primaries and secondaries.

Riparian Refers to streamside habitats. In Northern California, common riparian tree species are willows *(Salix)* and alders *(Alnus)*. Riparian habitat supports a higher diversity of nesting birds than any other habitat in North America and is also important for migrant land birds.

Roost A place where birds gather; also, to gather at such a place.

Scapulars Feathers located at the shoulder, sometimes creating a colorful patch.

Scavenger Any organism that feeds on carrion or organic refuse; ravens, crows, jays, and many gull species are scavengers.

Sculpins A group of small fish that inhabit intertidal zones and kelp forests.

Secondaries The inner flight feathers of a bird's wing.

Sibling species Species that are extremely similar in appearance but are reproductively isolated from one another. Sibling species are often thought to be relatively recent products of the speciation process (e.g., scaup).

Spatulate With a rounded, flattened shape; shovel shaped.

Split Refers to the division of what was thought to be a single species into two or more species by taxonomists (e.g., Western Grebe was split into Western and Clark's Grebes on the basis of demographic and ecological studies); opposite of lumped.

Superspecies Species that diverged from one another in isolation rather recently, and have remained largely or entirely geographically separated. Taxonomists group such closely related species that are allopatric (i.e., with nonoverlapping distributions) into superspecies.

Swale A depression in a hillside or low-lying area that holds soil moisture and supports wetland vegetation. Swales are a favored habitat of the Common Yellowthroat.

Sympatric Overlapping in geographical distribution.

Tarsus The longest, usually unfeathered, portion of a bird's leg, between the ankle (or tibia, which seems like the knee from a human perspective) and the foot. Plural is tarsi.

Taxonomy The practice of naming and classifying animals in order of evolutionary relationships.

Torpor A physiological condition in which the metabolism and body temperature are lowered to conserve energy during periods of environmental stress (cold). Hummingbirds use this survival strategy.

Tule Tall wetland plant of fresh and brackish marshes commonly associated with cattails; also known as bulrush.

Wrack Debris left along the high-water line of beaches and estuaries; often consists of shore-cast kelp, eelgrass, carcasses, and shells; an important source of nutrients, often picked through by shorebirds, ravens, gulls, and mammalian scavengers.

Xeric Refers to a dry environment.

Zygodactyl Refers to the anatomy of the avian foot in which two toes are forward and two (the first and fourth) are oriented backward, an adaptive design characteristic of parrots, woodpeckers, cuckoos, and the Osprey.

REFERENCES

Materials and Web Pages

Ainley, D. 1995. Ashy Storm-Petrel *(Oceanodroma homochroa)*, The Birds of North America Online (A. Poole, ed.). Ithaca, NY: Cornell Lab of Ornithology; retrieved from the Birds of North America Online: http://bna.birds.cornell.edu/bna/species/185.

Ainley, D., and R. Boekelheide, eds. 1990. *Seabirds of the Farallon Islands: Ecology, Dynamics, and Structure of an Upwelling-system Community*. Stanford, CA: Stanford University Press.

Ainley, D. G., and K. D. Hyrenbach 2010. "Top-Down and Bottom-Up Factors Affecting Seabird Population Trends in the California Current System (1985–2006)." *Progress in Oceanography* 84 (3–4): 242–254.

Anderson, D., D. Fix, K. Foerster, R. Hewitt, J. Hewston, T. Leskiw, G. S. Lester, R. LeValley, S. McAllister, and J. Power. 1995. *A Guide to Birding in and around Arcata*. City of Arcata, CA.

AOU (American Ornithologists' Union). 1985. "Thirty-fifth Supplement to the Check-list of North American Birds." *Auk* 102: 680–686.

AOU (American Ornithologists' Union). 2010. "Fifty-first Supplement to the American Ornithologists' Union Check-list of North American Birds." *Auk* 127 (3): 726–744.

AOU (American Ornithologists' Union). 2011. "Fifty-second Supplement to the Check-list of North American Birds." *Auk* 128: 600–613.

Beedy, E. C., and W. J. Hamilton, III. 1999. Tricolored Blackbird *(Agelaius tricolor)*, The Birds of North America Online (A. Poole, ed.). Ithaca, NY: Cornell Lab of Ornithology; retrieved from the

Birds of North America Online: http://bna.birds.cornell.edu/bna/species/423.

Carboneras, C. 1992. "Family Gaviidae (Divers)." In *Handbook of the Birds of the World, Volume 1: Ostrich to Ducks*, edited by Josep del Hoyo, Andrew Elliott, and Jordi Sargatal, 162–172. Barcelona: Lynx Edicions.

Cody, M. L. 2012. California Thrasher *(Toxostoma redivivum)*, The Birds of North America Online (A. Poole, Ed.). Ithaca, NY: Cornell Lab of Ornithology; retrieved from the Birds of North America Online: http://bna.birds.cornell.edu/bna/species/323.

Cracraft, J. 1983. "Species Concepts and Speciation Analysis." *Current Ornithology* 1: 159–187.

Dawson, W. L. 1923. *The Birds of California*. Volumes 1–4. San Diego, Los Angeles, San Francisco: South Moulton Company.

Evens, J., and I. Tait. 2005. *Introduction to California Birdlife*. California Natural History Guide Series No. 83. Berkeley: University of California Press.

Gaston, A. J., and S. B. Dechesne. 1996. Rhinoceros Auklet *(Cerorhinca monocerata)*, The Birds of North America Online (A. Poole, ed.). Ithaca, NY: Cornell Lab of Ornithology; retrieved from the Birds of North America Online: http://bna.birds.cornell.edu/bna/species/212.

Grinnell, J., H. C. Bryant, and T. I. Storer. 1918. *The Game Birds of California*. Berkeley: University of California Press.

Grinnell, J., and A. H. Miller. 1944. *The Distribution of the Birds of California*. Pacific Coast Avifauna No. 27. Berkeley, CA: Cooper Ornithological Society.

Gutierrez, R. J., M. Cody, S. Courtney, and A. B. Franklin. 2007. "The Invasion of Barred Owls and Its Potential Effect on the Spotted Owl: A Conservation Conundrum." *Biological Invasions* 9: 181–196.

Hackett, S. J. et al. 2008. "A Phylogenomic Study of Birds Reveals Their Evolutionary History." *Science* 320 (5884): 1763–1768.

Hagelin, J. C., and I. L. Jones. 2007. "Bird Odors and Other Chemical Substances: A Defense Mechanism or Overlooked Mode of Intraspecific Communication?" *Auk* 124 (3): 741–761.

Harris, S. W. 2005. *Northwestern California Birds: A Guide to Status, Distribution, and Habitats of the Birds of Del Norte, Humboldt, Trinity, Northern Mendocino, and Western Siskiyou Counties, California*, 3rd ed. Klamath River, CA: Living Gold Press.

Hatch, S. A., and D. N. Nettleship. 1998. Northern Fulmar *(Fulmarus glacialis)*, The Birds of North America Online (A. Poole, ed.). Ithaca, NY: Cornell Lab of Ornithology; retrieved from the Birds of North America Online: http://bna.birds.cornell.edu/bna/species/361.

Healey, W. M., and J. W. Thomas. 1973. "Effects of Dusting on Plumage of Japanese Quail." *Wilson Bulletin* 85: 442–448.

Hoffman, R. 1927. *Birds of the Pacific States.* Illustrations by Allan Brooks. Cambridge, MA: Houghton Mifflin, The Riverside Press.

Howell, S. N. G. 1999. "A Basic Understanding of Moult: What, When, Why, and How Much?" *Birders Journal* 8 (6): 296–300.

Howell, S. N. G. 2010. *Molt in North American Birds*. Boston, New York: Houghton Mifflin Harcourt.

Huntington, C. E., R. G. Butler, and R. A. Mauck. 1996. Leach's Storm-Petrel *(Oceanodroma leucorhoa)*, The Birds of North America Online (A. Poole, ed.). Ithaca, NY: Cornell Lab of Ornithology; retrieved from the Birds of North America Online: http://bna.birds.cornell.edu/bna/species/233.

Lowther, P. E. 1975. "Geographic and Ecological Variation in the Family Icteridae." *Wilson Bulletin* 87 (4): 481–495.

Manthey, J. D., J. Klicka, and G. M. Spellman. 2011. "Isolation Driven Divergence: Speciation in a Widespread North American Songbird (Aves: Certhiidae)." *Molecular Phylogenetic Ecology* 58 (3): 502–512.

Marshall, D. B., M. G. Hunter, and A. L Contreras. 2003, 2006. *Birds of Oregon: A General Reference*. Corvallis: Oregon State University Press.

Mayr, E. 1942. *Systematics and the Origin of Species from the Viewpoint of a Zoologist*. New York: Columbia University Press.

McShane, C., T. Hamer, H. Carter, G. Swartzman, V. Friesen, D. Ainley, R. Tressler, K. Nelson, A. Burger, L. Spear, T. Mohagen, R. Martin, L. Henkel, K. Prindle, C. Strong, and J. Keany. 2004. Evaluation Report for the 5-year Status Review of the Marbled Murrelet in Washington, Oregon, and California. Unpublished report by EDAW, Inc., Seattle, WA, prepared for the U.S. Fish and Wildlife Service, Region 1, Portland, Oregon.

Myers, J. P., P. G. Connors, and F. A. Pitelka. 1979. "Territoriality in Non-Breeding Shorebirds." *Studies in Avian Biology* 2: 231–232.

Nelson, S. K. 1997. Marbled Murrelet *(Brachyramphus marmoratus)*, The Birds of North America Online (A. Poole, ed.).

Ithaca, NY: Cornell Lab of Ornithology; retrieved from the Birds of North America Online: http://bna.birds.cornell.edu/bna/species/276.

Orians, G. H., and G. Collier. 1963. "Competition and Blackbird Social Systems." *Evolution* 17: 449–459.

Piersma, T. 1995. "Family Scolopacidae." In *The Handbook of Birds of the World*, vol. 3, 444–487. Barcelona: Lynx Edicions.

Poole, A. F., R. O. Bierregaard, and M. S. Martell. 2002. Osprey *(Pandion haliaetus)*, The Birds of North America Online (A. Poole, ed.). Ithaca, NY: Cornell Lab of Ornithology; retrieved from the Birds of North America Online: http://bna.birds.cornell.edu/bna/species/683.

Powell, A. N. 2001. "Habitat Characteristics and Nest Success of Snowy Plovers Associated with California Least Tern Colonies." *Condor* 103: 785–792.

Pyle, P. 1997. *Identification Guide to North American Birds*, part 1. Bolinas, CA: Slate Creek Press.

Pyle, P. 2005. "Molts and Plumages of Ducks (*natinae*)." *Waterbirds* 28: 208–209.

Pyle, P. 2008. *Identification Guide to North American Birds*. Part II. Slate Creek Press, Point Reyes Station, California. 835 pp.

Pyle, P. 2012. "In Memory of the Avian Species." *Birding* 44 (5): 34–39.

Reed, A., D. H. Ward, D. V. Derksen, and J. S. Sedinger. 1998. Brant *(Branta bernicla)*, The Birds of North America Online (A. Poole, ed.). Ithaca, NY: Cornell Lab of Ornithology; retrieved from the Birds of North America Online: http://bna.birds.cornell.edu/bna/species/337.

Reinking, D., and S. N. G. Howell. 1993. "An Arctic Loon in California." *Western Birds* 24: 189–196.

Roberson, D. 2002. *Monterey Birds: Status and Distribution of Birds in Monterey County, California*, 2nd ed. Carmel, CA: Monterey Peninsula Audubon Society.

Robinson, J. A., L. W. Oring, J. P. Skorupa, and R. Boettcher. 1997. American Avocet *(Recurvirostra americana)*, The Birds of North America Online (A. Poole, ed.). Ithaca, NY: Cornell Lab of Ornithology; retrieved from the Birds of North America Online: http://bna.birds.cornell.edu/bna/species/275.

Sauer, J. R., S. Schwartz, and B. Hoover. 1996. The Christmas Bird Count Home Page. Version 95.1. Patuxent Wildlife Research Center, Laurel, MD: www.mbr-pwrc.usgs.gov/bbs/cbc.html.

Shuford, W. D., C. M. Hickey, R. J. Safran, and G. Page. 1996. "A Review of the Status of the White-faced Ibis in California." *Western Birds* 27 (4): 169–196.

Shuford, W. D., and T. Gardali, eds. 2008. *California Bird Species of Special Concern: A Ranked Assessment of Species, Subspecies, and Distinct Populations of Birds of Immediate Conservation Concern in California*. Studies of Western Birds 1. Camarillo, CA: Western Field Ornithologists, and Sacramento: California Department of Fish and Game.

Stallcup, R. 1986. "Storm-Petrels." Focus in *Point Reyes Bird Observatory Newsletter* 15: 6–7.

Stallcup, R. 1990. *Ocean Birds of the Nearshore Pacific*. Stinson Beach, CA: Point Reyes Bird Observatory.

Storer, R. W., and G. L. Nuechterlein. 1992. Clark's Grebe *(Aechmophorus clarkii)*, The Birds of North America Online (A. Poole, ed.). Ithaca, NY: Cornell Lab of Ornithology; retrieved from the Birds of North America Online: http://bna.birds.cornell.edu/bna/species/026bdoi:10.2173/bna.26b.

Warner, R. E., and K. M. Hendrix. 1984. *Riparian Systems: Ecology, Conservation, and Productive Management*. Berkeley: University of California Press.

Websites

American Birding Association, *ABA Principles of Birding Ethics*. www.aba.org/bigday/ethics.pdf

American Ornithologist's Union Check-list of North American Birds. http://aou.org/checklist/north/print.php

Birds of North America Online. http://bna.birds.cornell.edu/bna

City-Data.com. www.city-data.com/states/California-Location-size-and-extent.html

National Marine Sanctuaries. http://sanctuaries.noaa.gov/

National Wildlife Refuge System. www.fws.gov/refuges/

Point Reyes Bird Observatory Conservation Science (aka Point Blue Conservation Science). www.pointblue.org

Proper Use of Playback in Birding. www.sibleyguides.com/2011/04/the-proper-use-of-playback-in-birding/

Ramsar Convention on Wetlands. www.ramsar.org

Redwood Regional Audubon Society. www.rras.org

Santa Cruz Bird Club. http://scbirdingguide.org

Sibley Guide: Identification of North American Birds and Trees. www.sibleyguides.com/

SORA. Searchable Ornithological Research Archive. http://elibrary.unm.edu/sora/
U.S. Fish and Wildlife Service. www.fws.gov/oregonfwo/Species/Data/NorthernSpottedOwl/BarredOwl/default.asp
Western Field Ornithologists. Official California Checklist by the California Bird Records Committee. Updated 3 August 2012. www.californiabirds.org/ca_list.asp
Western Hemisphere Shorebird Reserve Network. www.whsrn.org/about-whsrn; www.whsrn.org/western-hemisphere-shorebird-reserve-network; www.whsrn.org/site-profile/san-francisco-bay

Plate 1. Shorebirds on fall migration: Left and right front, a **Western Sandpiper** and a **Least Sandpiper**; left and right middle, a **Ruff** and a **Lesser Yellowlegs**; and rear, a **Red-necked Phalarope**. *August 1989. R.S.*

Plate 2. **Black-bellied Plover** is gray and white during winter but may be seen in fancy breeding plumage in April when on its way to the Arctic, or during July and August as it returns. *R.S.*

Plate 3. **American Avocet** in breeding plumage is one of the most striking of North American wading birds in shallow coastal embayments in the southern reaches of our region. *27 March 2011. Novato. J.E.*

Plate 4. Groups of small white sandpipers often seen on ocean beaches running in and out in the wave wash zone are **Sanderlings**. *R.S.*

Plate 5. **Black Oystercatcher** spends most days in the splash zone of the rocky shoreline; individuals occupy the same territory through their long lives (10 to 16 years). *1 January 2007. J.E.*

Plate 6. **California Brown Pelican** is a truly nearshore marine species, rarely venturing inland or very far out to sea. *20 September 2011. Bodega Bay. J.E.*

Plate 7. Close inspection of **Canada Goose** reveals wide variation in size and plumage of the "white-cheeked goose complex," with seven subspecies currently assigned among the large "honkers." *22 January 2013. Benicia. J.E.*

Plate 8. **Common Loon** is a common winter presence in Northern California, often found close to shore, foraging actively in the calmer waters of large embayments. *1 April 2011. Tomales Bay. J.E.*

Plate 9. This picture, taken in early April, shows the **Common Loon** undergoing molt in preparation for his spring migration to his northern breeding grounds. *1 April 2011. Tomales Bay. J.E.*

Plate 10. Not only do both species of **Dowitcher** look alike, but so do the sexes (monomorphism). Note the bill: it has sensitive nerve endings for detecting buried prey and a flexible tip to pick food from the substrate. *25 December 2009. Len Blumin.*

Plate 11. This **Long-billed Curlew's** bill is not exceptionally long for the species, suggesting that this is either a male or a younger bird. *2 March 2010. Tomales Bay. J.E.*

Plate 12. **Common Murre** is penguin-like in structure and coloration, but unlike penguins, murres can fly and arose from an entirely separate evolutionary lineage, a striking example of convergent evolution. *June 1977. Southeast Farallon Islands. J.E.*

Plate 13. **Black Brant**, a smallish sea goose, is an eelgrass specialist in winter, favoring estuaries where that plant grows abundantly below the water's surface. The bird pictured here has a blade of eelgrass dangling from its bill. *30 March 2013. Tomales Bay. J.E.*

Plate 14. **Aleutian Cackling Goose** is one of the four subspecies that might be encountered in winter along the coast. Cacklers often intermingle with much larger Canada Geese, especially in midwinter at traditional sites. *22 January 2013. Marin County. J.E.*

Plate 15. Although skittish around nesting sites, **Wood Ducks** may frequent city parks and ponds in the nonbreeding season. Here the male is on the left, female on the right. *5 February 2013. J.E.*

Plate 16. The drake **Gadwall's** plumage (foreground) shows a fine reticulated pattern. The female (behind) wears more subdued dress, similar to the hen Mallard, though slightly smaller and with a thinner bill. *5 February 2013. Marin County. J.E.*

Plate 17. The cinnamon flanks of **American Wigeon** are conspicuous on both the male (shown here) and the female. These surface-feeding ducks primarily eat aquatic plants but sometimes forage in flocks on upland grasses and forbs. *5 February 2013. J.E.*

Plate 18. The female **Mallard**, pictured here, has plumage similar to that of the female Gadwall, but the distinctive coloration of the bill and the sturdy structure help to distinguish her. *5 February 2013. J.E.*

Plate 19. **Blue-winged Teal** (male and female shown here) are unusual on the coast but may be found occasionally mixed in with Cinnamon Teal at marshy ponds, especially in early spring. *R.S.*

Plate 20. Few dabbling ducks are as striking as a drake **Cinnamon Teal** in full regalia, sporting powder-blue wing coverts and an iridescent green speculum that flashes in flight. *3 December 2009. Len Blumin.*

Plate 21. **Northern Shovelers** have large, orange, spatulate bills that may be difficult to see when the birds are foraging, straining invertebrates and seeds from the water as shown here. *R.S.*

Plate 22. The **Northern Pintail** female (foreground) is noticeably smaller than the male (behind). The profile of this species—long-necked, long-tailed, and rather elongated slender body—is unique among the dabblers. *16 December 2011. Marin County. J.E.*

Plate 23. The green chevron behind the eye of the male **Green-winged Teal** may remind you of the American Wigeon, but this is a much smaller dabbler with a rusty forehead and gray flanks. *23 November 2012. Len Blumin.*

Plate 24. This **Canvasback** (male), with the characteristic straight slope of the forehead, has just surfaced from a dive and has the bottom mud through which he was foraging coating his bill and face. *15 February 2013. J.E.*

Plate 25. **Redhead** numbers in California are low when prairie pothole nesting grounds are afflicted with drought and loss of habitat. Still, the occasional individual can be found amid Canvasback flocks. *10 December 2012. Marin County. J.E.*

Plate 26. When encountering a pair of waterfowl, it is advisable not to assume they are the same species. Here, a male **Ring-necked Duck** (left) is accompanied by a female **Lesser Scaup** (right). *6 January 2007. Marin County. J.E.*

Plate 27. **Surf Scoters** are abundant all along our coast during the winter, especially along the shore and in larger bays. Here, several males compete with each other to win the favor of females. *R.S.*

Plate 28. The least common of the three species of scoter, the **Black Scoter** is most often found in roiling water at headlands or sea stacks, but sometimes inside the larger bays. *R.S.*

Plate 29. The male **Bufflehead** uses his tail feathers for balance and leverage before leaping forward to dive. The female plumage is mostly gray brown with a small, distinctive white patch behind and below the eye. *6 January 2007. J.E.*

Plate 30. Pictured here are two male **Barrow's Goldeneyes** and one **Common Goldeneye**. The latter species far outnumbers the former in our bays and estuaries. *R.S.*

Plate 31. **Common Mergansers**, pictured here with a male in the center flanked by female-plumaged birds, are sturdy divers partial to freshwater lakes and streams. *R.S.*

Plate 32. **Red-breasted Merganser**, the most marine of our three species, has a more slender bill, neck, and body shape than the similar Common Merganser. The crest on this individual, slicked back from its recent dive, would be more prominent with drier plumage. *26 February 2012. Tomales Bay. J.E.*

Plate 33. The male **Ruddy Duck's** breeding plumage is radically different from his nonbreeding plumage. The upper parts molt to shiny cinnamon, and the crown to black, and the bill changes from dull gray to lazuli blue. *R.S.*

Plate 34. The female **Ruddy Duck** differs from a winter-plumaged male by the presence of a muddy-brown stripe through her white cheek. *R.S.*

Plate 35. This midwinter **Common Loon** is quite plain, lacking white spangles that will appear on the back in spring (see plates 8 and 9). Note the squared-off head and the pattern of a partial collar on the side of the neck. *26 February 2012. Tomales Bay. J.E.*

Plate 36. These larger grebes—Western and Red-necked—are somewhat similar in structure, but notice the head shapes: the **Western Grebe's** (foreground) head is more oval, the **Red-necked Grebe's** (background) head more triangular. *18 December 2009. Tomales Bay. J.E.*

Plate 37. A female **Pied-billed Grebe** is shown here with her zebra-faced youngsters. Pied-billeds prefer freshwater lakes and ponds during the nesting season, but may be found in small numbers on brackish estuarine waters in fall and winter. *R.S.*

Plate 38. The smaller grebes are similar, but their proportions differ. This **Eared Grebe** has a finer, more pointed bill and the stern seems to ride higher in the water, more buoyantly than the Horned Grebe. *16 January 2011. J.E.*

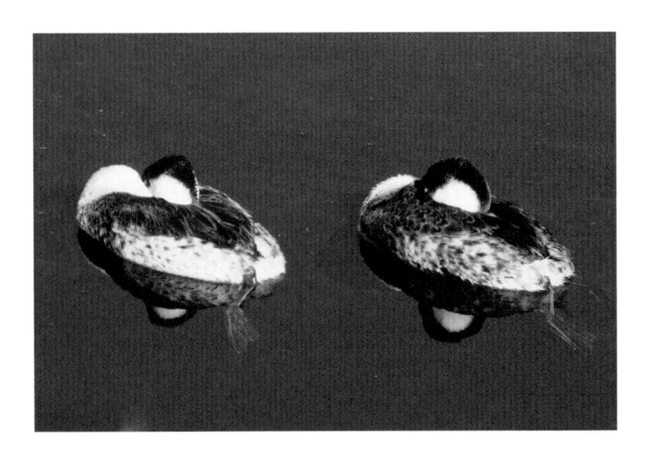

Plate 39. These two rafting individuals have their bills tucked beneath the dorsal feathering, so identification using bill color is not possible, but the darkness of the back feathers suggests **Western Grebe** rather than Clark's Grebe. *R.S.*

Plate 40. **Black-footed Albatross** are present throughout the year in offshore waters. Some individuals store nutrients in the crop while foraging near the California coast to deliver to their chicks in Hawaii. *R.S.*

Plate 41. This dark-plumaged **Northern Fulmar** shows an example of the unique bill structure that characterizes the "tubenoses." *22 September 2011. Cordell Bank. J.E.*

Plate 42. **Pink-footed Shearwaters**, though rarely seen from shore, may be abundant within 5 mi of the coast, especially in fall. *R.S.*

Plate 43. **Fork-tailed Storm-Petrel** is representative of the storm-petrels genus *Oceanodroma.* Note the tubed nose and the hooked beak, characteristic of all the family group. *20 September 2011. Cordell Bank. J.E.*

Plate 44. Most often seen roosting on pilings, rafts, and piers, the **Double-crested Cormorant** is the most common cormorant in estuaries and bays. *R.S.*

Plate 45. In *The Log from the Sea of Cortez* (1951), when John Steinbeck said, "Pelicans always seem to know where they are going," he was referring to Brown Pelicans, but the same could be said of **American White Pelicans**, pictured here. *26 January 2012. J.E.*

Plate 46. **Brown Pelicans** are long-lived birds and take three to five years to reach full adult plumage. Young birds do not have the white head of the adult, but have a more uniform brown plumage overall. *20 September 2011. Bodega Harbor, Sonoma County. J.E.*

Plate 47. **American Bitterns** rarely venture into the open, preferring to stay hidden in the thick stands of cattails and tules. When stalking prey, the head is often held high before striking. *R.S.*

Plate 48. This adult **Great Blue Heron** sports long, filamentous feathers on his breast (pectoral plumes) and, when foraging, has the intense stare characteristic of this species. *5 February 2013. J.E.*

Plate 49. This adult **Great Egret** is wading knee deep through aquatic vegetation, actively hunting for invertebrates, amphibians, fish, even small birds or mammals. *22 July 2011. Abbott's Lagoon, Point Reyes. J.E.*

Plate 50. This **Snowy Egret** is in full breeding plumage. Those elegant plumes nearly caused the species' demise when nearly all were killed so their plumes (aigrettes) would serve human vanity to decorate the headdresses of women. *27 March 2011. J.E.*

Plate 51. **Green Heron** is a stalker of riversides and pond edges, often standing statue still in wait for a small fish, frog, or crawfish. *12 December 2012. Lagunitas Creek, Marin County. J.E.*

Plate 52. **Black-crowned Night-Heron** (adult), while large, is shorter necked and stockier than the other ardeids. Note the large red eye and the stout bill. Shore crabs are favored prey. *14 January 2013. Sonoma County. J.E.*

Plate 53. With a 6 ft wing span, the **Turkey Vulture** soars over the coastal slope searching for carrion. Once food is detected—by the rising smell of the gases of decomposition—the vulture circles and descends. *11 December 2011. Point Reyes. J.E.*

Plate 54. When foraging on the wing, the **Osprey** or "fish hawk" flaps, glides, and hovers high above the water with head cocked forward, searching visually. *2 April 2011. Tomales Bay. J.E.*

Plate 55. Its hovering behavior has earned the **White-tailed Kite** the colloquial name "angel hawk." *April 2006. Marin County. Len Blumin.*

Plate 56. This **Northern Harrier** is an adult female. When hunting, harriers sail buoyant and low over meadows or marshes, ready to quarter and drop on any prey that moves or is flushed. *18 December 2008. Len Blumin.*

Plate 57. The fierce expression of this young **Cooper's Hawk** is characteristic of the species. The similar Sharp-shinned Hawk has a gentler expression, a smaller head, and a relatively shorter tail. *30 November 2011. Marin County. J.E.*

Plate 58. This **Red-shouldered Hawk** still retains the barred breast of a young bird but is beginning to develop some characteristics of an adult—rusty shoulders and black-and-white checkering on the upper wing. *24 January 2012. Marin County. J.E.*

Plate 59. The classic profile of our most common raptor, the **Red-tailed Hawk**, does not always show the rusty tail as it is viewed from below, especially when it is backlighted or an immature bird. *1 December 2011. Marin County. J.E.*

Plate 60. This adult **Golden Eagle** shows the golden nape for which this iconic raptor is named. Most birds seen along the immediate coast are subadults wandering from interior nesting sites, unattached to a territory. *R.S.*

Plate 61. The **American Coot** is an atypical member of the Rallidae family. Rather than skulking around in marsh vegetation, coots are easily observed, gregarious, gathering in large flocks and foraging on open water. *J. E.*

Plate 62. The **Black Rail** avoids sunlight for the dark corridors of the tidal marsh, where it scurries through runways within the pickleweed forest, seeming to avoid flight at all costs. *10 March 2005. San Pablo Bay. Danika Tsao.*

Plate 63. A **"California" Clapper Rail** forages along the bank of a tidal slough. As the water rises, Clapper Rails take refuge in tidal marsh vegetation of the upper marsh plain. *27 February 2013. San Francisco Bay. J.E.*

Plate 64. The **Virginia Rail** may venture out from the cover of the marsh vegetation to forage along the shoreline but is always ready to scamper back into the safety of the cattails. *R.S.*

Plate 65. The **American Coot** often leaves the water to graze on grasses. Note the large feet with lobed toes, adapted to walking on aquatic vegetation and paddling underwater. *16 December 2011. J.E.*

Plate 66. The name of the **Black-bellied Plover** refers to this striking breeding plumage, which is attained in late March or April. Adult plovers may still be in this plumage when they first return in late summer or early fall. *29 April 2009. Len Blumin.*

Plate 67. Early in the nesting season, **Western Snowy Plover** adult males are distinguished from females by the rusty cap and the black patches on the crown, ear, and sides of the neck. Females are generally duller, with brown flecking in amid the black markings. *12 April 2011. Point Reyes. J.E.*

Plate 68. Smaller than the Killdeer and larger than the Snowy Plover, the **Semipalmated Plover** is not an uncommon sight on the outer beach or tidal flat, especially during migratory periods. *2 March 2013. Point Reyes. J.E.*

Plate 69. The fledgling **Killdeer**, pictured here with an adult, has only one black necklace. Confusion with Semipalmated Plover is a possible identification trap, but the Killdeer hatchling is much darker above. *R.S.*

Plate 70. **Black Oystercatchers** are in their element—dense beds of bivalves. Contrary to the species' name, oysters are rarely taken—mussels, limpets, and crabs are the main food items. *12 November 2012. Kehoe Beach, Point Reyes. J.E.*

Plate 71. The aptly named **Black-necked Stilt**, with its extraordinary long legs, is closely related to avocets, but its bill is very slightly upturned. *17 November 2009. Marin County. J.E.*

Plate 72. **American Avocet** in breeding plumage, with the soft rusty wash to its head and neck, is one of the most striking of North American wading birds. *27 March 2011. Marin County. J.E.*

Plate 73. **Sanderlings** wear a whitish plumage in winter and are the most common small sandpiper on the outer beach. *28 October 2011. Drake's Beach, Point Reyes. J.E.*

Plate 74. The **Spotted Sandpiper** is not very spotted in basic winter plumage, but you may spot one, usually singly, along a calm shoreline. *28 February 2013. Marin County. J.E.*

Plate 75. A typical representative of the tringine tribe of sandpipers, the **Greater Yellowlegs** is a fairly common shorebird in both freshwater and tidal habitats, where the water is shallow. *5 September 2011. J.E.*

Plate 76. The **Willet** on the left is in the drab, basic plumage most often seen along these shores in winter. In this April photograph, the Willet on the right is coming into the more patterned breeding dress. *10 April 2011. Tomales Bay. J.E.*

Plate 77. The **Long-billed Curlew** (left) has a warmer cinnamon feathering than the **Whimbrel** (middle) but is very similar in color to the **Marbled Godwit** (right). Note the differing bill shapes. *11 December 2011. San Francisco Bay. J.E.*

Plate 78. The long, upcurved bill of the **Marbled Godwit**—pink at the base, dark toward the tip—is used to extract polychaete worms and small crustaceans from tidal flats along the California coast. *30 March 2013. Tomales Bay. J.E.*

Plate 79. **Black Turnstone**—black above, white below—is a stocky bird with short legs and a stout bill. In flight, the wings flash a striking black-and-white pattern. Flocks are quite vocal, chattering with harsh rattles and cackles. *2 March 2013. Drake's Beach, Point Reyes. J.E.*

Plate 80. **Surfbirds** are well named—sturdy sandpipers that have evolved the abilities to forage the outermost surf-battered rocks agilely, avoiding the crashing waves. *R.S*

Plate 81. In basic nonbreeding plumage, worn for most of the year along these shores, **Sanderlings** are the lightest colored of our small shorebirds, blending in with the sandy beaches they usually frequent. Note the characteristic pot-bellied shape. *3 March 2013. Drake's Beach, Point Reyes. J.E.*

Plate 82. *Peeps* is a term used to describe small sandpipers similar in structure, shape, and size. The two larger birds in the middle are **Western Sandpipers**. The four smaller birds are **Least Sandpipers**. *13 September 2011. Tomales Bay, Marin County. J.E.*

Plate 83. **Dunlin**—larger than the peeps with which it consorts—is one of the most abundant small shorebirds visiting these shores in winter, especially the larger estuaries. *3 February 2004. Len Blumin.*

Plate 84. **Dowitchers** tend to gather in small flocks at the water's edge. The two species are so similar, especially in basic plumage (shown here), that vocalizations are the most reliable clue to identity. *8 November 2010. J.E.*

Plate 85. **Wilson's Snipe**, unlike most other shorebirds, rarely flies in flocks, although a preference for marshy substrate concentrates numbers of them into the same habitat patches. *R.S.*

Plate 86. This first-cycle **Red-necked Phalarope** pauses to forage during fall migration, en route from the Arctic to the Southern Hemisphere. Remarkably, these delicate shorebirds spend about three-quarters of their lives at sea. *R.S.*

Plate 87. The **Common Murre** is one of our most abundant nesting seabirds, penguin-like in structure and coloration, the Northern Hemisphere equivalent of that Southern Hemisphere icon. *25 June 2011. J.E.*

Plate 88. This adult **Pigeon Guillemot** is in breeding plumage. The origin of *guillemot* is the Celtic word *geolaff,* meaning "weep," which may be imitative of the sound of the bird's sibilant vocalizations. *R.S.*

Plate 89. This pair of **Marbled Murrelets** is in breeding plumage. This small seabird is difficult to see from shore, especially in summer, because of its cryptic sooty plumage and its habit of frequent, prolonged dives. *7 July 2010. Humboldt County. Sean McAllister.*

Plate 90. **Cassin's Auklet** chicks emerge from the egg with a full coat of fluffy down. The chick will leave the nesting burrow for the ocean at about six weeks of age and forage independent of adults. *June 1977. Southeast Farallon Islands. J.E.*

Plate 91. This adult **Tufted Puffin** is in full breeding regalia. The winter plumage is much darker, absent the white face, the streaming plumes, and the bright red bill. *R.S.*

Plate 92. This adult **Heermann's Gull**, a handsome gull in any plumage, is in definitive basic plumage here. In alternate (breeding) plumage, attained in winter, the head will lose the gray streaking and become entirely white. *20 September 2011. Bodega Bay. J.E.*

Plate 93. This adult **Mew Gull** in nonbreeding winter plumage has an overall delicate structure and relatively straight, slender bill with a subtle gonydeal angle to the lower mandible. *13 January 2013. Sausalito, Marin County. J.E.*

Plate 94. This adult **Ring-billed Gull** is in basic winter plumage. The yellow eye, surrounded by a narrow red eye-ring, placed rather forward on the face gives this gull a seemingly studious look. *10 December 2012. Sausalito, Marin County. J.E.*

Plate 95. This adult **California Gull** in November shows basic winter plumage. Note the rather dark blue-gray mantle, the leg color, and the bicolored bill with a red blotch behind the black on the lower mandible. *27 November 2011. Marin County. J.E.*

Plate 96. The adult **Thayer's Gull** in basic winter plumage closely resembles the Herring Gull, but with a seemingly gentler facial expression because of a rounder head and less massive bill. *6 January 2007. J.E.*

Plate 97. This group of large, white-headed gulls has gathered at a herring spawn in San Francisco Bay. Three species are shown—**Western Gull**, **Glaucous-winged Gull**, and **Herring Gull** in the center. *21 January 2013. Sausalito, Marin County. J.E.*

Plate 98. A young **Caspian Tern** (left) begs for food from an adult (right). These birds traveled more than 20 mi from the nearest nesting site in San Francisco Bay to forage on the outer coast. *22 July 2011. Abbott's Lagoon. J.E.*

Plate 99. Dust bathing is a common behavior in **California Quail**, pictured here, as well as in other gallinaceous birds (grouse, pheasants, turkeys) and also in Mourning Doves and House Sparrows, among others. *25 January 2013. Point Reyes. J.E.*

Plate 100. The **Mourning Dove** is fairly common along the coastal strip. The body shape is typical of the family, although this species is rather slender. The light-blue eye-ring indicates this is an adult. *14 March 2013. J.E.*

Plate 101. **Burrowing Owl**, a grassland ground nester of the interior, visits the coastal counties from Sonoma County south, but sparingly, in the nonnesting season. It is a California Bird Species of Special Concern. *15 April 2011. Novato, Marin County. J.E.*

Plate 102. The **Northern Spotted Owl**, like this adult in coastal redwood, is the true and gentle spirit of the mature coastal conifer forest, north of the Golden Gate. *21 March 2010. Napa County. J.E.*

Plate 103. The **Black Swift**, one of our rarest and most enigmatic aerial insectivores, nests behind or within the spray zone of waterfalls and on coastal headlands. *July 2013. Eric Horvath.*

Plate 104. **Allen's Hummingbird** (male, here), a bold and brilliant bantam, is noticeably smaller than Anna's Hummingbird. The flight display of the male is a remarkable and extravagant aerial ballet. *Len Blumin.*

Plate 105. **Acorn Woodpeckers** are habitual hoarders, storing and husbanding acorns for future consumption. The pockets excavated to store the acorns do not reach the cambium and do not harm the tree. *25 January 2013. Bear Valley, Point Reyes. J.E.*

Plate 106. A male **American Kestrel**, a sit-and-wait predator, searching for prey—grasshoppers, beetles, mice and other small rodents; even lizards may be taken. *R.S.*

Plate 107. An adult male **Merlin** roosted in oak above a wetland observes a distant flock of shorebirds. Note the compact shape, the heavily streaked underparts, and the bluish dorsal coloration of a male. *5 December 2011. Novato, Marin County. J.E.*

Plate 108. The heavy vertical streaking on its breast indicates that this Peale's **Peregrine Falcon** is a young bird. An adult would have finer horizontal barring across the breast. Note the large talons. *22 March 2013. Novato, Marin County. J.E.*

Plate 109. **Pacific-slope Flycatcher**, formerly called Western Flycatcher, belongs to the notoriously difficult to identify *Empidonax* complex. This species is by far the commonest member of the group along our coast. *15 September 2012. Outer Point Reyes. J.E.*

Plate 110. Although not closely related, **Hutton's Vireo** (right) and **Ruby-crowned Kinglet** (left) are superficially similar in appearance; the vireo is larger headed with a stouter bill and thicker legs, and it is not "flitty" in the field. *R.S.*

Plate 111. **Common Ravens** pay close attention to other animals and are adept at exploiting every opportunity. The massive bill helps distinguish this corvid from the much smaller American Crow. *J.E.*

Plate 112. **Tree Swallow** is perhaps the most common of the five common swallow species encountered on the coast. The male (pictured here) is a brilliant metallic blue above and immaculate white below, the female somewhat duller. *21 April 2013. Point Reyes Station. J.E.*

Plate 113. **Chestnut-backed Chickadee** is a sassy year-round resident along the coast, foraging busily for insects among needles, leaves, and bark, though small seeds are also favored and suburban seed feeders are visited. *24 March 2013. Point Reyes Station. J.E.*

Plate 114. The **Pacific Wren**, shorter tailed and chunkier than our other wrens, is a bird of the interior damp forest, usually glimpsed skulking through the undergrowth of ferns and brambles. *13 October 2012. Point Reyes. J.E.*

Plate 115. The **Wrentit's** long tail and short wings are adaptations to life in the tangled shrubbery in which it spends its life. *R.S.*

Plate 116. **Varied Thrush** is similar in size and shape to the more familiar American Robin, but with striking orange patterns on the face, wings, and breast. *5 February 2013. J.E.*

Plate 117. **Northern Mockingbirds** often find a conspicuous perch from which to advertise their impressive repertoire. They have expanded their range dramatically with the suburbanization of coastal California but are more common in the southern counties. *19 March 2013. J.E.*

Plate 118. **American Pipit** has gray-brown upper parts and a (variably) streaked breast with a slight buffy wash. In nonbreeding plumage it may at first be mistaken for a thrush. *9 February 2006. Limantour Beach. Len Blumin.*

Plate 119. **Common Yellowthroat** is a wetland warbler. The black mask is distinctive in the male, absent in the female. Her crown and cheek have a more subdued olive-brown wash. *Ian Tait.*

Plate 120. **Song Sparrow** is resident year-round. Although fairly widely distributed, its greatest abundance is associated with wetlands—especially riparian corridors and marshes. Fox, Lincoln's, and Savannah Sparrows are similar in appearance. *R.S.*

Plate 121. **Savannah Sparrow** is similar in general impression to the Song Sparrow but often shows yellow in the facial pattern and, overall, a lighter dorsal coloration than the local Song Sparrows. *6 December 2012. Point Reyes. J.E.*

Plate 122. **"Nuttall's" White-crowned Sparrow** is a coastal race of the species, a resident of maritime chaparral and coastal sage scrub in which it sings its clarion song. *11 January 2012. Point Reyes. J.E.*

Plate 123. The **Dark-eyed Junco**, a year-round resident of coastal counties, is one of the more common "yard birds," especially when flocks are swelled by wintering birds. *7 Marsh 2013. Point Reyes Station. J.E.*

Plate 124. **Western Meadowlark** is a ground forager, eating seeds and insects, as well as a ground nester. Like many other flocking birds, it flashes white outer tail feathers in flight. *1 March 2013. Outer Point Reyes. J.E.*

Plate 125. This **Brown-headed Cowbird**, newly fledged, possibly came from the nest of a White-crowned Sparrow or Yellow Warbler. Note the stubby, conical bill, the expressionless face, and the relatively short wings. *22 June 2012. J.E.*

Plate 126. This **Red-winged Blackbird**, the bicolored version, is displaying and erecting the bright red epaulets for which it is named. *5 May 2006. Len Blumin.*

Plate 127. The male **Tricolored Blackbird** is easily identified by the red-and-white shoulder patch, although both colors are not always visible, as shown here. *January 2005. R.S.*

Plate 128. The male **Brewer's Blackbird** has shiny black plumage and a piercing yellow eye. The female, shown here, is dressed in drabber charcoal gray, with a reddish-brown eye. *15 February 2013. J.E.*

Plate 129. This male **Lesser Scaup** shows the typical head shape, bill pattern, and head shine that helps separate the species from the Greater Scaup. *November 1982. R.S.*

Plate 130. **Spotted Towhee** (male, here) is a year-round resident of coastal scrub, forest edge, and residential gardens. It is highly territorial and monogamous. A catlike "meow" is usually the first clue to its presence. *30 March 2013. J.E.*

Plate 131. In rural coastal settings, **Purple Finch** (male, here) is a common member of the nesting community, although the species can be hard to find during its winter wanderings. *7 March 2013. J.E.*

Plate 132. **Lesser Goldfinch** (male left, female right) is a western species, especially common in California. Its diet consists almost entirely of the small seeds of plants in the family Compositae. *7 March 2013. J.E.*

Plate 133. This female **Belted Kingfisher** is holding a newt in her bill. The diverse diet of kingfishers is revealed by the compact pellets that adults disgorge from their perches. These pellets contain bones, scales, and other indigestible remains of their prey. Photo by Ian Tait.

INDEX

Page numbers in **bold** indicate the main discussion of the taxon. Page numbers followed by f indicate figures; plates are indicated by pl. followed by the plate number.

ABOUT THE AUTHORS

Rich Stallcup (1944–2012) was a preeminent California field ornithologist, naturalist, and conservationist. He was founder of the Point Reyes Bird Observatory and the author of many articles and books on bird identification, biogeography, and conservation.

Jules Evens (1947–) is a wildlife biologist with four decades of experience observing Northern California's coastal birdlife. His previous books include *The Natural History of the Point Reyes Peninsula* (UC Press, 2008) and *An Introduction to California Birdlife* (UC Press, 2005).

California Natural History Guides

From deserts and grasslands to glaciers and the spectacular Pacific coast, California is a naturalist's paradise. The authors of the California Natural History Guides have walked into wildfires, plunged into shark-infested waters, scaled the Sierra Nevada, peeked under rocks, and gazed into the sky to present the most extensive California environmental education series in existence. Packed with photographs and illustrations and compact enough to take on the trail, they are essential reading for any California adventurer. For a complete list of guides, please visit http://www.ucpress.edu/go/cnhg.